Springer Tracts in Natural Philosophy

Volume 21

Edited by B. D. Coleman

Elna B. McBride

Obtaining Generating Functions

Springer-Verlag New York Heidelberg Berlin 1971

Elna Browning McBride
Professor of Mathematics
Memphis State University
Memphis, Tennessee 38111

AMS Subject Classifications (1970):
33-00, 33-02

ISBN 0-387-05255-0 Springer-Verlag New York Heidelberg Berlin
ISBN 3-540-05255-0 Springer-Verlag Berlin Heidelberg New York

 Library of Congress Catalog Card Number 72-138 811. Printed in Germany. Typesetting, printing, and binding: Universitätsdruckerei H. Stürtz AG, Würzburg.

Preface

This book is an introduction to the study of methods of obtaining generating functions. It is an expository work at the level of the beginning graduate student.

The first part of Chapter I gives the reader the necessary definitions and basic concepts. The fundamental method of direct summation is explained and illustrated.

The second part of Chapter I deals with the methods developed by Rainville. These methods are based principally on inventive manipulation of power series.

Weisner's group-theoretic method is explained in detail in Chapter II and is further illustrated in Chapter III. When this method is applicable, it yields a set of at least three generating functions. In Chapter II for the Laguerre polynomials six generating functions were found.

Truesdell's method is studied in Chapter IV. For a given set of functions $\{f(z, \alpha)\}$ the success of this method depends on the existence of certain transformations. If $f(z, \alpha)$ can be transformed into $F(z, \alpha)$ such that

$$\frac{\partial}{\partial z} F(z, \alpha) = F(z, \alpha+1),$$

or if $f(z, \alpha)$ can be transformed into $G(z, \alpha)$ such that

$$\frac{\partial}{\partial z} G(z, \alpha) = G(z, \alpha-1),$$

then from each transformed function a generating function can be obtained. Truesdell's method for obtaining the transformed functions does not require any ingenuity on the user's part. Truesdell has shown how these simple results may be exploited to generate more complicated results by means of specified, systematic, and general processes. His method of obtaining generating functions is only one of these results.

Although the principal objective of this exposition is to bring to the reader's attention the methods developed by Rainville, Truesdell, and

Weisner, there are other methods in the literature which deserve consideration. Some of these are presented in Chapter V.

The author is especially grateful that Professor Truesdell, Professor Weisner, and the late Professor Rainville read and made valuable suggestions concerning the parts of the original manuscript dealing with the method developed by each of them.

Contents

Chapter I

Series Manipulation Methods

First Part. Underlying Ideas

1. Introduction. The purpose of this study is to describe and to make illustrative use of some effective methods for obtaining generating functions. We define a generating function for a set of functions $\{f_n(x)\}$ as follows: Let $G(x,t)$ be a function that can be expanded in powers of t such that

$$G(x,t)=\sum_{n=0}^{\infty} c_n f_n(x)\, t^n,$$

where c_n is a function of n that may contain the parameters of the set $\{f_n(x)\}$, but is independent of x and t. Then $G(x,t)$ is called a generating function of the set $\{f_n(x)\}$.

To illustrate we generate the set of functions $\{1, x, x^2, \ldots, x^n, \ldots\}$. We know that

$$\exp\{x\,t\}=\sum_{n=0}^{\infty}(x\,t)^n/n!=\sum_{n=0}^{\infty}\frac{1}{n!}x^n t^n.$$

Then corresponding to the notation in our definition of a generating function we have

$$G(x,t)=\exp\{x\,t\}, \quad c_n=1/n!, \quad \text{and} \quad f_n(x)=x^n.$$

By the above definition a set of functions may have more than one generating function. However, if

$$G(x,t)=\sum_{n=0}^{\infty} h_n(x)\, t^n$$

then $G(x,t)$ is the unique generator for the set $\{h_n(x)\}$ *as the coefficient set*. For example, the set of functions $\{x^n\}$ is generated as a coefficient set only by $(1-x\,t)^{-1}$.

We use the symbol $\{f_n(x)\}$ to indicate the infinite set $\{f_0(x), f_1(x), f_2(x), \ldots, f_n(x), \ldots\}$. If $f_n(x)$ is also defined for negative integral n, we would like to find a function $H(x,t)$ having a Laurent series expansion of the form

$$H(x,t) = \sum_{n=-\infty}^{\infty} c_n f_n(x) t^n.$$

Presently, we will extend our definition of generating function to include functions whose expansions are Laurent series.

We define a *formal* power series as one for which the radius of convergence is not necessarily greater than zero. When a function $H(x,t)$ has a power series expansion in t, then $H(x,t)$ determines the coefficient set $\{h_n(x)\}$ even if the series is divergent for $t \neq 0$. The relation between the generating function and the coefficient set is a qualitative relation whose validity does not depend on the length of the radius of convergence. In 1923 Eric T. Bell [1] presented a paper in which he established the validity of "results obtained by equating coefficients after formal manipulation of series". (Also see Bell [2] and [3].) Accordingly, we do not consider it necessary to determine the radius of convergence for the power series representation of each generating function. However, if the generating function has a power series expansion which is obviously divergent for $t \neq 0$, we will use the following notation to indicate divergence:

$$H(x,t) \cong \sum_{n=0}^{\infty} h_n(x) t^n.$$

We now extend our definition of a generating function to include functions with a Laurent series expansion, functions whose expansions have a zero radius of convergence, and finally functions which generate functions of more than one variable. (See Erdélyi [3; p. 228].) Let $G(x_1, x_2, \ldots, x_p; t)$ be a function of $p+1$ variables. Suppose

$$G(x_1, x_2, \ldots, x_p; t)$$

has a formal expansion in powers of t such that

$$G(x_1, x_2, \ldots, x_p; t) = \sum_{n=-\infty}^{\infty} c_n f_n(x_1, x_2, \ldots, x_p) t^n$$

where c_n is independent of the variables $x_1, x_2, \ldots, x_p$, and t. Then we shall say that $G(x_1, x_2, \ldots, x_p; t)$ is a generating function for the $f_n(x_1, x_2, \ldots, x_p)$ corresponding to nonzero c_n. In particular, if

$$G(x,y;t) = \sum_{n=0}^{\infty} c_n f_n(x) g_n(y) t^n,$$

the expansion determines the set of constants $\{c_n\}$ and the two sets of functions $\{f_n(x)\}$ and $\{g_n(y)\}$. Then $G(x, y; t)$ is to be considered as a generator of any one of these three sets and as the unique generator of the coefficient set $\{c_n f_n(x) g_n(y)\}$.

A generating function may be used to define a set of functions, to determine a differential recurrence relation or a pure recurrence relation, to evaluate certain integrals, etc. We will use generating functions to define the following special functions: the Bessel functions and the polynomials of Legendre, Gegenbauer, Hermite, and Laguerre.

The Legendre polynomials $\{P_n(x)\}$ were introduced by Legendre [1] in 1785. He defined them by means of the generating relation:

$$(1-2xt+t^2)^{-\frac{1}{2}} = \sum_{n=0}^{\infty} P_n(x)\, t^n. \tag{1}$$

In 1874 Gegenbauer [1; pp. 6—16] generalized the Legendre polynomials and used the notation $\{C_n^\nu(x)\}$ for the set which satisfies the generating relation:

$$(1-2xt+t^2)^{-\nu} = \sum_{n=0}^{\infty} C_n^\nu(x)\, t^n. \tag{2}$$

These polynomials are now called the Gegenbauer polynomials. In this book we adopt Legendre's definition of $P_n(x)$ and Gegenbauer's definition of $C_n^\nu(x)$.

We define the Hermite polynomials $\{H_n(x)\}$ by means of the generating relation

$$\exp\{2xt-t^2\} = \sum_{n=0}^{\infty} \frac{1}{n!} H_n(x)\, t^n. \tag{3}$$

When Hermite [1; p. 294] introduced these polynomials in 1864, he used the symbol U_n and defined U_n by what we call a Rodrigues-type relation:

$$e^{-x^2} U_n = \frac{d^n}{dx^n} e^{-x^2}.$$

The functions $H_n(x)$ and U_n differ only in sign, i.e., $H_n(x) = (-1)^n U_n$. However, the twentieth century notation is not uniform. Magnus and Oberhettinger [1; p. 80] use a different symbol, $He_n(x)$, to warn us of a different definition:

$$\exp\{xt-t^2/2\} = \sum_{n=0}^{\infty} \frac{1}{n!} He_n(x)\, t^n. \tag{4}$$

If in (4) we replace t by $t\sqrt{2}$ and x by $x\sqrt{2}$, we get

$$\exp\{2xt-t^2\} = \sum_{n=0}^{\infty} \frac{1}{n!} [He_n(x\sqrt{2})](t\sqrt{2})^n. \tag{5}$$

By comparing coefficients of t in (5) and (3) we see that

$$H_n(x)=2^{n/2}\,He_n(x\sqrt{2}).$$

Erdélyi [2; p. 192] lists a number of authors and specifies the notation used by each.

We define the Laguerre polynomials $\{L_n^{(\alpha)}(x)\}$ by means of the generating relation

$$(1-t)^{-1-\alpha}\exp\left\{\frac{-x\,t}{1-t}\right\}=\sum_{n=0}^{\infty}L_n^{(\alpha)}(x)\,t^n. \tag{6}$$

If $\alpha=0$, these polynomials are denoted by $\{L_n(x)\}$ and are called the simple Laguerre polynomials. The polynomials which Laguerre [1; p. 430] actually introduced were the simple Laguerre. He used the symbol $f_m(x)$ and defined this set $\{f_m(x)\}$ by means of the differential equation

$$x\,y''+(x+1)\,y'-m\,y=0. \tag{7}$$

The Laguerre polynomial $L_m^{(\alpha)}(x)$, as we have defined it, satisfies the differential equation

$$x\,y''+(1-\alpha-x)\,y'+m\,y=0. \tag{8}$$

By comparing (7) and (8) we see that

$$f_m(x)=L_m(-x).$$

Even in the case of the simple Laguerre polynomials, the notation in current literature is not uniform. For example, Sneddon [1; p. 160] defines the (simple) Laguerre polynomials by means of the generating relation

$$\exp\left\{\frac{-x\,t}{1-t}\right\}=(1-t)\sum_{n=0}^{\infty}\frac{1}{n!}L_n(x)\,t^n.$$

If we set $\alpha=0$ in (6) above, we see that our $L_n(x)$ is equivalent to Sneddon's $L_n(x)/n!$.

We define the class of modified Laguerre polynomials as that class whose elements are sets of the form $\{b(n)\,L_n^{\alpha(n)}(x)\}$, where $b(n)$ and $\alpha(n)$ are functions of n independent of x. If $b(n)=1$ and $\alpha(n)=0$, we have the simple Laguerre polynomials $\{L_n(x)\}$. If $b(n)=1$ and $\alpha(n)=\alpha$, where α is a nonnegative constant, we have the (generalized) Laguerre polynomials $\{L_n^{\alpha}(x)\}$. If $b(n)=(-1)^n$ and $\alpha(n)=-\beta-n$, where β is a constant, we have the set $\{(-1)^n L_n^{-\beta-n}(x)\}$, which we will use frequently. Let $f_n^{\beta}(x)\equiv(-1)^n L_n^{-\beta-n}(x)$. The set $\{f_n^{\beta}(x)\}$ satisfies the generating relation

$$(1-t)^{-\beta}\,e^{x\,t}=\sum_{n=0}^{\infty}f_n^{\beta}(x)\,t^n. \tag{9}$$

We now define the set of Bessel functions; simple Bessel polynomials will be defined in Chapter III, Section 3. The Bessel functions $\{J_n(x)\}$, for integral n and for $t \neq 0$, satisfy the generating relation

$$\exp\left\{\frac{x}{2}(t-t^{-1})\right\} = \sum_{n=-\infty}^{\infty} J_n(x)\, t^n. \tag{10}$$

These functions as we have defined them (with n an integer) are sometimes called Bessel coefficients. In 1824 Bessel [1; pp. 92 and 100] defined I_k^h and J_k^h by means of the following integrals, where k is the independent variable:

$$2\pi\, I_k^h = \int_0^{2\pi} \cos(h\,\varepsilon - k\,\sin\varepsilon)\, d\varepsilon$$

and

$$2\pi\, J_k^h = \int_0^{2\pi} \frac{\cos(h\,\varepsilon - k\,\sin\varepsilon)\, d\varepsilon}{1-\varepsilon\,\cos\varepsilon}.$$

The $J_n(x)$ of our definition (10) has the integral representation

$$\pi\, J_n(x) = \int_0^{\pi} \cos(n\,\varepsilon - x\,\sin\varepsilon)\, d\varepsilon.$$

See Rainville [1; p. 114] and Whittaker and Watson [1; p. 362]. Also, $J_n(x)$ has the series representation

$$J_n(x) = \sum_{k=0}^{\infty} \frac{(-1)^k (x/2)^{n+2k}}{k!\,(n+k)!}. \tag{11}$$

In accordance with present-day usage we will use the symbol $I_n(x)$ to designate the following function:

$$I_n(x) = i^{-n} J_n(i\,x) = \sum_{k=0}^{\infty} \frac{(x/2)^{n+2k}}{k!\,(n+k)!}. \tag{12}$$

The function $J_n(x)$ is described as a Bessel function of the first kind of index n, and $I_n(x)$ is called a modified Bessel function of the first kind of index n. The function $I_n(x)$ is also referred to as the hyperbolic Bessel function. Whittaker and Watson [1; p. 373] list various relationships for $I_n(z)$ including an integral representation. For a discussion of $J_n(z)$ when n is not necessarily an integer see Erdélyi [2; pp. 1–114], Watson [1], Whittaker and Watson [1; pp. 355–385].

For a given set of functions it is desirable to have as many generating functions as possible from which to choose the one best suited for a particular use. The methods for obtaining generating functions which are discussed in greatest detail in this study are Weisner's group theoretic method, Truesdell's F-equation method, and methods depending on series manipulation.

We will use series manipulation as the fundamental method in Chapter I. Also series manipulative techniques are auxiliary to some of the methods of Chapter V. The method of series manipulation depends upon the following basic relations:

$$\sum_{n=0}^{\infty}\sum_{k=0}^{\infty} A(k,n) = \sum_{n=0}^{\infty}\sum_{k=0}^{n} A(k,n-k) \tag{13}$$

and

$$\sum_{n=0}^{\infty}\sum_{k=0}^{\infty} A(k,n) = \sum_{n=0}^{\infty}\sum_{k=0}^{[n/2]} A(k,n-2k). \tag{14}$$

We use the symbol $[n/2]$ to denote the greatest integer less than or equal to $n/2$. The proofs of (13) and (14) are found in Rainville [1; pp. 56–58]. From (13) and (14) it follows that

$$\sum_{n=0}^{\infty}\sum_{k=0}^{n} C(k,n) = \sum_{n=0}^{\infty}\sum_{k=0}^{[n/2]} C(k,n-k). \tag{15}$$

With the aid of (14) we will now illustrate for the Hermite polynomials a procedure leading from a given generating function to a series representation. Assume as given the defining generating relation

$$\exp\{2xt-t^2\} = \sum_{n=0}^{\infty}\frac{1}{n!}H_n(x)\,t^n.$$

We expand the generating function as follows:

$$\begin{aligned}\exp\{2xt-t^2\} &= \exp\{2xt\}\exp\{-t^2\}\\ &= \sum_{n=0}^{\infty}\frac{(2xt)^n}{n!}\sum_{k=0}^{\infty}\frac{(-t^2)^k}{k!}.\end{aligned}$$

Be means of (14) we have

$$\begin{aligned}\exp\{2xt-t^2\} &= \sum_{n=0}^{\infty}\sum_{k=0}^{[n/2]}\frac{(2xt)^{n-2k}}{(n-2k)!}\frac{(-t^2)^k}{k!}\\ &= \sum_{n=0}^{\infty}\sum_{k=0}^{[n/2]}\frac{(-1)^k(2x)^{n-2k}}{(n-2k)!\,k!}t^n.\end{aligned}$$

By comparing coefficients of t^n in this result and in our defining generating relation, we get

$$\frac{H_n(x)}{n!} = \sum_{k=0}^{[n/2]}\frac{(-1)^k(2x)^{n-2k}}{(n-2k)!\,k!}.$$

Before attempting a similar derivation for any other set, we will define the very useful factorial function $(a)_n$.

2. The factorial function and the generalized hypergeometric functions. We define the factorial function $(a)_n$ as follows: For any number a

$$(a)_n = a(a+1)(a+2)\cdots(a+n-1), \qquad \text{for } n \geqq 1,$$

and

$$(a)_0 = 1 \qquad \text{for } a \neq 0.$$

The factorial function is an extension of the ordinary factorial since $(1)_n = n!$. On the basis of the definition of the gamma function, we may write

$$(a)_n = \frac{\Gamma(a+n)}{\Gamma(a)},$$

if a is neither zero nor a negative integer.

We will now establish some factorial function identities which will be used in this chapter. By regrouping factors we may express $(a)_{2n}$ as follows:

$$\begin{aligned}(a)_{2n} &= [a(a+2)\cdots(a+2n-2)]\,[(a+1)(a+3)\cdots(a+2n-1)] \\ &= 2^{2n}\left(\frac{a}{2}\right)_n \left(\frac{a+1}{2}\right)_n.\end{aligned} \tag{1}$$

If in identity (1) we let $a=1$, we get

$$(2n)! = 2^{2n}(\tfrac{1}{2})_n\, n!. \tag{2}$$

By introducing factors in both numerator and denominator of $(a)_{n-k}$ as defined we get

$$\begin{aligned}(a)_{n-k} &= \frac{a(a+1)\cdots(a+n-k-1)\,[(a+n-k)\cdots(a+n-1)]}{[(a+n-k)\cdots(a+n-1)]} \\ &= \frac{(a)_n}{(-1)^k(1-a-n)_k}.\end{aligned} \tag{3}$$

If in identity (3) we let $a=1$, we get

$$(n-k)! = \frac{n!}{(-1)^k(-n)_k}. \tag{4}$$

It is particularly convenient to use the factorial function when indicating a binomial expansion:

$$\begin{aligned}(1-t)^{-a} &= \sum_{n=0}^{\infty} \frac{(-a)(-a-1)\cdots(-a-n+1)}{n!}(-t)^n \\ &= \sum_{n=0}^{\infty} \frac{(a)_n}{n!}\, t^n.\end{aligned} \tag{5}$$

The usefulness of this notation is evident in the following derivation of a series representation of the Gegenbauer polynomial $C_n^\nu(x)$ from the generating relation

$$(1-2xt+t^2)^{-\nu}=\sum_{n=0}^{\infty} C_n^\nu(x)\,t^n.$$

We expand the generating function of this relation by means of the binomial expansion:

$$(1-2xt+t^2)^{-\nu}=[1-(2xt-t^2)]^{-\nu}=\sum_{n=0}^{\infty}(\nu)_n\frac{(2xt-t^2)^n}{n!}.$$

Again we use the factorial function form of the binomial expansion to change the form of the expression on the right:

$$\begin{aligned}(1-2xt+t^2)^{-\nu}&=\sum_{n=0}^{\infty}\frac{(\nu)_n}{n!}(2xt)^n(1-t/2x)^n\\&=\sum_{n=0}^{\infty}\frac{(\nu)_n}{n!}(2xt)^n\sum_{k=0}^{n}\frac{(-n)_k}{k!}\left(\frac{t}{2x}\right)^k.\end{aligned}$$

By applying (15) of Section (1), we obtain

$$(1-2xt+t^2)^{-\nu}=\sum_{n=0}^{\infty}\sum_{k=0}^{[n/2]}\frac{(\nu)_{n-k}(2x)^{n-2k}(-1)^k}{(n-2k)!\,k!}\,t^n.$$

Finally, by equating coefficients of t^n, we prove that

$$C_n^\nu(x)=\sum_{k=0}^{[n/2]}\frac{(-1)^k(\nu)_{n-k}(2x)^{n-2k}}{(n-2k)!\,k!}. \tag{6}$$

In a similar manner we will obtain from the defining relation

$$(1-t)^{-1-\alpha}\exp\left\{\frac{-xt}{1-t}\right\}=\sum_{n=0}^{\infty}L_n^{(\alpha)}(x)\,t^n$$

a series representation for the Laguerre polynomial $L_n^{(\alpha)}(x)$. The given generating function may be written as follows:

$$\begin{aligned}(1-t)^{-1-\alpha}\exp\left\{\frac{-xt}{1-t}\right\}&=(1-t)^{-1-\alpha}\sum_{k=0}^{\infty}\frac{1}{k!}\left(\frac{-xt}{1-t}\right)^k\\&=\sum_{k=0}^{\infty}\frac{(-xt)^k}{k!}(1-t)^{-1-k-\alpha}.\end{aligned}$$

By expanding the binomial in the expression on the right, we get

$$(1-t)^{-1-\alpha}\exp\left\{\frac{-xt}{1-t}\right\}=\sum_{k=0}^{\infty}\frac{(-xt)^k}{k!}\sum_{n=0}^{\infty}\frac{(1+k+\alpha)_n}{n!}t^n$$
$$=\sum_{n=0}^{\infty}\sum_{k=0}^{n}\frac{(-x)^k(1+\alpha)_n}{k!(n-k)!(1+\alpha)_k}t^n.$$

Therefore, by equating coefficients of t^n, we find

$$L_n^{(\alpha)}(x)=\sum_{k=0}^{n}\frac{(-1)^k(1+\alpha)_n x^k}{k!(n-k)!(1+\alpha)_k}. \tag{7}$$

For later reference we also include a series derivation for the modified Laguerre polynomial $f_n^{\beta}(x)$, where $f_n^{\beta}(x)=(-1)^n L_n^{-\beta-n}(x)$, the derivation being based on the given relation

$$e^{xt}(1-t)^{-\beta}=\sum_{n=0}^{\infty}f_n^{\beta}(x)\,t^n.$$

The left member of this relation is expanded as follows:

$$e^{xt}(1-t)^{-\beta}=\sum_{n=0}^{\infty}\frac{(xt)^n}{n!}\sum_{k=0}^{\infty}\frac{(\beta)_k}{k!}t^k$$
$$=\sum_{n=0}^{\infty}\sum_{k=0}^{n}\frac{(\beta)_k x^{n-k}}{(n-k)!\,k!}t^n.$$

In order to change x^{n-k} to x^k we commute the terms of this finite series by interchanging k and $n-k$. It then follows that

$$f_n^{\beta}(x)=\sum_{k=0}^{n}\frac{(\beta)_k x^{n-k}}{(n-k)!\,k!}=\sum_{k=0}^{n}\frac{(\beta)_{n-k}x^k}{k!(n-k)!}. \tag{8}$$

We will have occasion to use frequently the hypergeometric function

$$F\left[\begin{matrix}a, & b; \\ & c;\end{matrix}\quad z\right]=\sum_{n=0}^{\infty}\frac{(a)_n(b)_n}{(c)_n}\frac{z^n}{n!}$$

and its generalization

$${}_pF_q\left[\begin{matrix}a_1, a_2, \dots, a_p; \\ b_1, b_2, \dots, b_q;\end{matrix}\quad z\right]=\sum_{n=0}^{\infty}\frac{(a_1)_n(a_2)_n\dots(a_p)_n}{(b_1)_n(b_2)_n\dots(b_q)_n}\frac{z^n}{n!}$$

where no denominator parameter can be zero or a negative integer. However, we do not require that p or q be different from zero. We illustrate the usefulness of the ${}_pF_q$ notation by using it to represent the exponential function, the binomial function, the Bessel function $J_n(x)$, the Laguerre polynomial $L_n^a(x)$, the Hermite polynomial $H_n(x)$, and the Gegenbauer polynomial $C_n^a(x)$. We assume that the expansion in powers of x for each function is known and convert it to the ${}_pF_q$ form.

$$e^x = \sum_{n=0}^{\infty} \frac{x^n}{n!} = {}_0F_0\left[\begin{matrix} -; \\ -; \end{matrix}\ x\right]. \tag{9}$$

$$(1-x)^{-a} = \sum_{n=0}^{\infty} \frac{(a)_n x^n}{n!} = {}_1F_0\left[\begin{matrix} a; \\ -; \end{matrix}\ x\right]. \tag{10}$$

$$\begin{aligned} J_n(x) &= \sum_{k=0}^{\infty} \frac{(-1)^k x^{n+2k}}{2^{n+2k} k!\,\Gamma(1+n+k)} \\ &= \frac{x^n}{2^n \Gamma(1+n)}\, {}_0F_1\left[\begin{matrix} -; \\ 1+n; \end{matrix}\ \frac{-x^2}{4}\right]. \end{aligned} \tag{11}$$

$$\begin{aligned} L_n^a(x) &= \sum_{k=0}^{n} \frac{(-1)^k (1+a)_n x^k}{k!\,(n-k)!\,(1+a)_k} \\ &= \frac{(1+a)_n}{n!}\, {}_1F_1\left[\begin{matrix} -n\,; \\ 1+a; \end{matrix}\ x\right]. \end{aligned} \tag{12}$$

$$\begin{aligned} H_n(x) &= \sum_{k=0}^{[n/2]} \frac{(-1)^k n!\,(2x)^{n-2k}}{k!\,(n-2k)!} \\ &= (2x)^n\, {}_2F_0\left[\begin{matrix} -\frac{n}{2}, \frac{-n+1}{2}; \\ -\ ; \end{matrix}\ -\frac{1}{x^2}\right]. \end{aligned} \tag{13}$$

$$\begin{aligned} C_n^a(x) &= \sum_{k=0}^{[n/2]} \frac{(-1)^k (a)_{n-k} (2x)^{n-2k}}{k!\,(n-2k)!} \\ &= \frac{(a)_n (2x)^n}{n!}\, {}_2F_1\left[\begin{matrix} \frac{-n}{2}, \frac{-n+1}{2}; \\ 1-a-n\ ; \end{matrix}\ \frac{1}{x^2}\right]. \end{aligned} \tag{14}$$

See Rainville [1; pp. 279–280, (15), (16), (20)] for other ${}_2F_1$ forms of $C_n^a(x)$.

3. Obtaining generating functions from expansions in powers of x. If a set of functions $\{f_n(x)\}$ is defined by means of a series representation of the form

$$f_n(x) = \sum_{k=0}^{n} F(k, x),$$

it is often possible to use series manipulation to find a generating function $G(x, t)$ such that

$$\sum_{n=0}^{\infty} g_n f_n(x)\, t^n = \sum_{n=0}^{\infty} g_n \left[\sum_{k=0}^{n} F(k, x) \right] t^n = G(x, t),$$

where g_n is not dependent on x and t.

An an illustration of this procedure, we use the simple Laguerre polynomials $\{L_n(x)\}$. We assume as given the series representation

$$L_n(x) = \sum_{k=0}^{n} \frac{(-1)^k\, n!\, x^k}{(k!)^2 (n-k)!}.$$

Since $n!$ is independent of k, we may also write

$$\frac{L_n(x)}{n!} = \sum_{k=0}^{n} \frac{(-1)^k x^k}{(k!)^2 (n-k)!}.$$

In this case we will find that it is possible to determine a generating function for each of the sets $\{L_n(x)\}$ and $\left\{\frac{L_n(x)}{n!}\right\}$.

A generating function for the set $\left\{\frac{L_n(x)}{n!}\right\}$ is found first. From (12) of Section 2 with $a=0$ we obtain a series representation of $L_n(x)$ which we use as follows:

$$\begin{aligned}\sum_{n=0}^{\infty} \frac{L_n(x)}{n!}\, t^n &= \sum_{n=0}^{\infty} \sum_{k=0}^{n} \frac{(-1)^k x^k}{(k!)^2 (n-k)!}\, t^n \\ &= \sum_{n=0}^{\infty} \sum_{k=0}^{\infty} \frac{(-1)^k x^k t^{n+k}}{(k!)^2\, n!}.\end{aligned}$$

The factors depending on n may be separated from those depending on k:

$$\sum_{n=0}^{\infty} \frac{L_n(x)}{n!}\, t^n = \sum_{n=0}^{\infty} \frac{t^n}{n!} \sum_{k=0}^{\infty} \frac{(-1)^k (x t)^k}{(k!)^2}.$$

Therefore, we have established the generating relation

$$\sum_{n=0}^{\infty} \frac{L_n(x)}{n!}\, t^n = e^t\, {}_0F_1\left[\begin{matrix} -; \\ 1; \end{matrix}\ -x t\right]. \tag{1}$$

See Rainville [1; p. 213, (3)].

The series manipulation procedure is slightly more difficult for the set $\{L_n(x)\}$:

$$\sum_{n=0}^{\infty} L_n(x)\, t^n = \sum_{n=0}^{\infty} \sum_{k=0}^{n} \frac{(-1)^k\, n!\, x^k}{(k!)^2 (n-k)!}\, t^n$$
$$= \sum_{n=0}^{\infty} \sum_{k=0}^{\infty} \frac{(-1)^k (n+k)!\, x^k\, t^{n+k}}{(k!)^2\, n!}.$$

By interchanging the order of summation, we get

$$\sum_{n=0}^{\infty} L_n(x)\, t^n = \sum_{k=0}^{\infty} \frac{(-1)^k (x\,t)^k}{k!} \sum_{n=0}^{\infty} \frac{(1+k)_n\, t^n}{n!}$$
$$= \sum_{k=0}^{\infty} \frac{(-x\,t)^k}{k!} (1-t)^{-1-k}.$$

Therefore, since $(1-t)^{-1}$ is independent of k, we have

$$\sum_{n=0}^{\infty} L_n(x)\, t^n = (1-t)^{-1} \exp\left\{\frac{-x\,t}{1-t}\right\}. \tag{2}$$

See Rainville [1; p. 213, (4a)].

For any given set $\{\phi_n(x)\}$ for which a series representation is known we may introduce a nonzero numerator parameter c and sometimes find a family of generating functions (one for each value of c):

$$G(x,t,c) = \sum_{n=0}^{\infty} (c)_n\, \phi_n(x)\, t^n.$$

As an illustration we use the set $\{f_n^{\beta}(x)\}$, where $f_n^{\beta}(x) = (-1)^n L_n^{-\beta-n}(x)$. Using the series representation of $f_n^{\beta}(x)$ given in (8) of Section 2, we write

$$\sum_{n=0}^{\infty} (c)_n\, f_n^{\beta}(x)\, t^n \cong \sum_{n=0}^{\infty} \sum_{k=0}^{n} \frac{(c)_n (\beta)_{n-k}\, x^k\, t^n}{(n-k)!\, k!}$$
$$\cong \sum_{n=0}^{\infty} \sum_{k=0}^{\infty} \frac{(c)_{n+k} (\beta)_n\, x^k\, t^{n+k}}{n!\, k!}.$$

The factors on the right may be rearranged as follows:

$$\sum_{n=0}^{\infty} (c)_n\, f_n^{\beta}(x)\, t^n \cong \sum_{n=0}^{\infty} \frac{(c)_n (\beta)_n\, t^n}{n!} \sum_{k=0}^{\infty} \frac{(c+n)_k (x\,t)^k}{k!}.$$

We have thus obtained the family of divergent generating functions

$$\sum_{n=0}^{\infty} (c)_n\, f_n^{\beta}(x)\, t^n \cong (1-x\,t)^{-c}\, {}_2F_0\left[\begin{matrix} c, \beta; \\ -\,; \end{matrix}\ \frac{t}{1-x\,t}\right]. \tag{3}$$

Second Part. Rainville's Methods

A generating function of the type obtained by the method of Section 4 is essential in the development of the methods of Sections 5 and 6.

4. Using an auxiliary variable. Suppose we have a generating function given for some set $\{f_n(x)\}$ and want to obtain one for the set $\{f_{n+k}(x)\}$, where k is a nonnegative integer. Let $G(x, t)$ represent the given generating function with the indicated expansion:

$$G(x,t)=\sum_{n=0}^{\infty} a_n f_n(x)\, t^n.$$

For this relation we seek a generalization:

$$K(x,t,k)=\sum_{n=0}^{\infty} b(n,k) f_{n+k}(x)\, t^n,$$

such that $K(x, t, 0)=G(x, t)$ and $b(n, 0)=a_n$.

The auxiliary variable v is introduced by replacing t by $t+v$ in the given generating relation:

$$G(x,t+v)=\sum_{n=0}^{\infty} a_n f_n(x)(t+v)^n.$$

Since

$$(t+v)^n=\sum_{k=0}^{n} \frac{n!\, t^{n-k}\, v^k}{(n-k)!\, k!},$$

we find that

$$\sum_{n=0}^{\infty} a_n f_n(x)(t+v)^n=\sum_{k=0}^{\infty}\sum_{n=0}^{\infty} \frac{(n+k)!\, a_{n+k} f_{n+k}(x)\, t^n}{n!\, k!}\, v^k.$$

If $G(x, t+v)$ can be expanded in powers of v in a different way, the desired relation can be found by equating coefficients of v^k.

We will illustrate this method by using two examples. First, we will generate the set $\{H_{n+k}(x)\}$, where $\{H_n(x)\}$ is the set of Hermite polynomials. We assume as given the generating relation:

$$\exp\{2xt-t^2\}=\sum_{n=0}^{\infty} \frac{H_n(x)}{n!}\, t^n. \tag{1}$$

In this relation we replace t by $t+v$:

$$\exp\{2x(t+v)-(t+v)^2\}=\sum_{n=0}^{\infty} \frac{H_n(x)}{n!}(t+v)^n. \tag{1a}$$

We now expand the left and right members of (1a) in powers of v in two different ways. After simplification the right member may be expressed as follows:

$$\sum_{n=0}^{\infty} \frac{H_n(x)}{n!}(t+v)^n = \sum_{k=0}^{\infty} v^k \sum_{n=0}^{\infty} \frac{H_{n+k}(x)\, t^n}{n!\, k!}.$$

Let $G(x,t)=\exp\{2xt-t^2)$. Then the left member of (1a) may be represented by $G(x,t+v)$. Before expanding $G(x,t+v)$ in powers of v we will associate the factors of this function which are independent of v:

$$G(x,t+v)=\exp\{2xt-t^2\}\exp\{2(x-t)v-v^2\}.$$

The factor which is dependent on v may then be expanded in powers of v by using (1):

$$\exp\{2(x-t)v-v^2\} = \sum_{k=0}^{\infty} \frac{H_k(x-t)}{k!} v^k.$$

Hence the left member of (1a) has the following expansion:

$$G(x,t+v)=\exp\{2xt-t^2\}\sum_{k=0}^{\infty} \frac{H_k(x-t)}{k!} v^k.$$

By equating coefficients of v^k in these two expansions we get the generating relation for the set $\{H_{n+k}(x)\}$:

$$\exp\{2xt-t^2\}\, H_k(x-t) = \sum_{n=0}^{\infty} \frac{H_{n+k}(x)}{n!} t^n. \tag{2}$$

See Rainville [1; p. 197, (1)], Truesdell [1; p. 85, (10)] and Weisner [2; p. 144, (4.3) with $w=0$].

As our second example we use $\{f_n^\beta(x)\}$, the set of modified Laguerre polynomials which satisfy the generating relation

$$e^{xt}(1-t)^{-\beta} = \sum_{n=0}^{\infty} f_n^\beta(x)\, t^n. \tag{3}$$

In order to introduce the auxiliary variable v we replace t by $t+v$:

$$\exp\{x(t+v)\}[1-(t+v)]^{-\beta} = \sum_{n=0}^{\infty} f_n^\beta(x)(t+v)^n. \tag{3a}$$

By expanding the binomial $(t+v)^n$ and simplifying we may write

$$\sum_{n=0}^{\infty} f_n^\beta(x)(t+v)^n = \sum_{k=0}^{\infty}\sum_{n=0}^{\infty} \frac{(n+k)!\, f_{k+n}^\beta(x)\, t^n}{n!\, k!} v^k.$$

We now try to expand the left member of (3a) in a different way.

If in the left member of (3a) we associate the factors depending on v, we get

$$\exp\{x(t+v)\}(1-t-v)^{-\beta}=e^{xt}(1-t)^{-\beta}\left[e^{xv}\left(1-\frac{v}{1-t}\right)^{-\beta}\right].$$

By means of (3) we may write

$$e^{xv}\left(1-\frac{v}{1-t}\right)^{-\beta}=\exp\left\{x(1-t)\frac{v}{1-t}\right\}\left(1-\frac{v}{1-t}\right)^{-\beta}$$
$$=\sum_{k=0}^{\infty} f_k^{\beta}(x(1-t))\left(\frac{v}{1-t}\right)^k.$$

Therefore, the left member of (3a) has the following expansion in powers of v:

$$\exp\{x(t+v)\}(1-t-v)^{-\beta}=e^{xt}(1-t)^{-\beta}\sum_{k=0}^{\infty} f_k^{\beta}(x(1-t))\left(\frac{v}{1-t}\right)^k.$$

Hence by equating coefficients of v^k in the two expansions, we get

$$\sum_{n=0}^{\infty}\frac{(n+k)!\, f_{k+n}^{\beta}(x)}{n!\,k!}t^n=e^{xt}(1-t)^{-\beta-k} f_k^{\beta}(x(1-t)). \tag{4}$$

The generating function of (4) is a special case of a generating function obtained in Chapter III, Section 2.

5. A bilinear generating function. If a function $G(x,y,t)$ can be expanded in the form

$$G(x,y,t)=\sum_{n=0}^{\infty} g_n f_n(x) f_n(y) t^n,$$

where g_n is independent of x and y, then $G(x,y,t)$ is called a bilinear generating function. For example, the Hermite polynomials satisfy the following bilinear generating relation:

$$\sum_{n=0}^{\infty}\frac{H_n(x)\,H_n(y)\,t^n}{n!}=(1-4t^2)^{-\frac{1}{2}}\exp\left\{y^2-\frac{(y-2xt)^2}{1-4t^2}\right\}. \tag{1}$$

See Rainville [1; p. 198, (2)] and Erdélyi [2; p. 194, (22), with $z=2t$].

If we extend the definition of a bilinear generating function so that we require only that

$$G(x,y,t)=\sum_{n=0}^{\infty} g_n f_{\alpha(n)}(x) f_{\beta(n)}(y) t^n,$$

where $\alpha(n)$ and $\beta(n)$ are functions of n which are not necessarily equal, then Weisner [2; p. 145, (4.9) with $y=2t$ and $w=y$] furnishes us with another example of a bilinear generating function which is a generalization of (1):

$$\sum_{n=0}^{\infty} \frac{H_n(x)\, H_{\nu+n}(y)}{n!}\, t^n = (1-4t^2)^{-(\nu+1)/2} \exp\left\{y^2 - \frac{(y-2xt)^2}{1-4t^2}\right\} H_\nu\left(\frac{y-2xt}{1-4t^2}\right). \tag{2}$$

The method for obtaining bilinear generating functions, which we explain and illustrate in this section, was used by Rainville [1; p. 197 and p. 211] to generate $\left\{\frac{H_n(x)\, H_n(y)}{n!}\right\}$ for the Hermite polynomials and $\left\{\frac{n!\, L_n^{(\alpha)}(x)\, L_n^{(\alpha)}(y)}{(1+\alpha)_n}\right\}$ for the Laguerre polynomials. For our first illustration let us consider

$$\sum_{n=0}^{\infty} \frac{(b)_n\, L_n^{(\alpha)}(x)\, L_n^{(\alpha)}(y)\, t^n}{(c)_n},$$

where b and c are to be chosen later for simplification purposes. (If we had started out without these parameters, the need for them would have forced us to come back and put them in.) We first replace $L_n^{(\alpha)}(x)$ by its expansion in powers of x:

$$\begin{aligned}
\sum_{n=0}^{\infty} \frac{(b)_n\, L_n^{(\alpha)}(x)\, L_n^{(\alpha)}(y)\, t^n}{(c)_n}
&= \sum_{n=0}^{\infty} \frac{(b)_n\, L_n^{(\alpha)}(y)\, t^n}{(c)_n} \sum_{k=0}^{n} \frac{(-1)^k (1+\alpha)_n\, x^k}{k!\,(n-k)!\,(1+\alpha)_k} \\
&= \sum_{n=0}^{\infty} \sum_{k=0}^{\infty} \frac{(b)_{n+k}\, L_{n+k}^{(\alpha)}(y)\, t^{n+k} (-1)^k (1+\alpha)_{n+k}\, x^k}{(c)_{n+k}\, k!\, n!\, (1+\alpha)_k}.
\end{aligned} \tag{3}$$

In order to use this method we must have available a generating function for the set $\{L_{n+k}^{(\alpha)}(y)\}$. (Methods for obtaining such generating functions are discussed in Chapter I, Section 4, in Chapter II, and in Chapter IV.) From Rainville [1; p. 211, (9)] we have

$$\sum_{n=0}^{\infty} \frac{(n+k)!\, L_{n+k}^{(\alpha)}(x)\, t^n}{k!\, n!} = (1-t)^{-1-\alpha-k} \exp\left\{\frac{-xt}{1-t}\right\} L_k^{(\alpha)}\left(\frac{x}{1-t}\right). \tag{4}$$

At this point we see that in order to use (4), we need in (3) the numerator factor $(n+k)!$ instead of the numerator factor $(1+\alpha)_{n+k}$. Accordingly, we choose $b=1$ and $c=1+\alpha$. We have on the basis of this choice

$$\sum_{n=0}^{\infty} \frac{n!\, L_n^{(\alpha)}(x)\, L_n^{(\alpha)}(y)\, t^n}{(1+\alpha)_n} = \sum_{k=0}^{\infty} \sum_{n=0}^{\infty} \frac{(n+k)!\, L_{n+k}^{(\alpha)}(y)\, t^n}{k!\, n!} \frac{(-xt)^k}{(1+\alpha)_k}. \tag{3a}$$

In (3a) we substitute the generating function of (4):

$$\sum_{n=0}^{\infty} \frac{n!\, L_n^{(\alpha)}(x)\, L_n^{(\alpha)}(y)\, t^n}{(1+\alpha)_n}$$

$$= \sum_{k=0}^{\infty} \left[(1-t)^{-1-\alpha-k} \exp\left\{\frac{-yt}{1-t}\right\} L_k^{(\alpha)}\left(\frac{y}{1-t}\right)\right] \frac{(-xt)^k}{(1+\alpha)_k} \tag{5}$$

$$= (1-t)^{-1-\alpha} \exp\left\{\frac{-yt}{1-t}\right\} \sum_{k=0}^{\infty} \frac{L_k^{(\alpha)}\left(\frac{y}{1-t}\right)}{(1+\alpha)_k} \left(\frac{-xt}{1-t}\right)^k.$$

In order to simplify this result we need the following generating relation which may be found by the basic summation method of Section 3:

$$\sum_{k=0}^{\infty} \frac{L_k^{(\alpha)}(x)}{(1+\alpha)_k} t^k = e^t\, {}_0F_1(-;1+\alpha;-xt). \tag{6}$$

(See Rainville [1; p. 201, (1)] or Erdélyi [2; p. 189, (18)].) If in (6) we replace x by $\frac{y}{1-t}$ and t by $\frac{-xt}{1-t}$, we have

$$\sum_{k=0}^{\infty} \frac{L_k^{(\alpha)}\left(\frac{y}{1-t}\right)}{(1+\alpha)_k} \left(\frac{-xt}{1-t}\right)^k = \exp\left\{\frac{-xt}{1-t}\right\} {}_0F_1\left[\begin{matrix} -; \\ 1+\alpha; \end{matrix} \; \frac{xyt}{(1-t)^2}\right]. \tag{7}$$

By substituting (7) in the final form of the right member of (5), we get

$$\sum_{n=0}^{\infty} \frac{n!\, L_n^{(\alpha)}(x)\, L_n^{(\alpha)}(y)\, t^n}{(1+\alpha)_n} = (1-t)^{-1-\alpha} \exp\left\{\frac{-t(x+y)}{1-t}\right\} {}_0F_1\left[\begin{matrix} -; \\ 1+\alpha; \end{matrix} \; \frac{xyt}{(1-t)^2}\right]. \tag{8}$$

(See Rainville [1; p. 212, Theorem 69].) This bilinear generating function appears in a different form in Erdélyi [2; p. 189, (20)].

As our second example we seek a bilinear generating function for the set $\{f_n^{\beta}(x)\}$, where

$$f_n^{\beta}(x) \equiv (-1)^n L_n^{-\beta-n}(x).$$

In order to find a bilinear generating function by the method of this section we need the series representation

$$f_n^{\beta}(x) = \sum_{k=0}^{n} \frac{(\beta)_{n-k}\, x^k}{(n-k)!\, k!} = \sum_{k=0}^{n} \frac{(\beta)_k\, x^{n-k}}{k!\,(n-k)!}, \tag{9}$$

the generating relation obtained in Article 4

$$\sum_{n=0}^{\infty} \frac{(n+k)!\, f_{k+n}^{\beta}(x)\, t^n}{k!\, n!} = e^{xt}(1-t)^{-\beta-k} f_k^{\beta}(x(1-t)), \tag{10}$$

and the (divergent) generating relation obtained in Article 3

$$\sum_{n=0}^{\infty} (c)_n f_n^{\beta}(x)\, t^n \cong (1-xt)^{-c}\, {}_2F_0\left[\begin{matrix} c, & \beta; & \dfrac{t}{1-xt} \\ & -; & \end{matrix}\right]. \tag{11}$$

We begin with

$$\sum_{n=0}^{\infty} n!\, f_n^{\beta}(x)\, f_n^{\beta}(y)\, t^n,$$

where $n!$ has been introduced to provide the $(n+k)!$ needed later. We replace $f_n^{\beta}(y)$ by its series representation given in (9):

$$\begin{aligned}\sum_{n=0}^{\infty} n!\, f_n^{\beta}(x) f_n^{\beta}(y)\, t^n &= \sum_{n=0}^{\infty} n!\, f_n^{\beta}(x)\, t^n \sum_{k=0}^{n} \frac{(\beta)_k\, y^{n-k}}{k!\,(n-k)!} \\ &= \sum_{n=0}^{\infty} \sum_{k=0}^{\infty} \frac{(n+k)!\, f_{n+k}^{\beta}(x)\, t^{n+k} (\beta)_k\, y^n}{k!\, n!}.\end{aligned}$$

We now substitute the generating function of (10) for the corresponding series:

$$\begin{aligned}\sum_{n=0}^{\infty} n!\, f_n^{\beta}(x) f_n^{\beta}(y)\, t^n &= \sum_{k=0}^{\infty} [e^{xyt}(1-yt)^{-k-\beta} f_k^{\beta}(x(1-yt))]\, (\beta)_k\, t^k \\ &= e^{xyt}(1-yt)^{-\beta} \sum_{k=0}^{\infty} (\beta)_k f_k^{\beta}(x(1-yt)) \left(\frac{t}{1-yt}\right)^k.\end{aligned}$$

But from (11) with $c=\beta$, we have

$$\sum_{k=0}^{\infty} (\beta)_k f_k^{\beta}(x(1-yt)) \left(\frac{t}{1-yt}\right)^k \cong (1-xt)^{-\beta}\, {}_2F_0\left[\begin{matrix} \beta, & \beta; & \dfrac{t}{(1-xt)(1-yt)} \\ - & ; & \end{matrix}\right].$$

Therefore, the appropriate substitution yields the following bilinear generating relation:

$$\begin{aligned}&\sum_{n=0}^{\infty} n!\, f_n^{\beta}(x) f_n^{\beta}(y)\, t^n \\ &\quad \cong e^{xyt}(1-xt)^{-\beta}(1-yt)^{-\beta}\, {}_2F_0\left[\begin{matrix} \beta, & \beta; & \dfrac{t}{(1-xt)(1-yt)} \\ - & ; & \end{matrix}\right].\end{aligned}$$

6. Bilateral generating functions. If $H(x, y, t)$ can be expanded in powers of t in the form

$$H(x, y, t) = \sum_{n=0}^{\infty} h_n f_n(x)\, g_n(y)\, t^n,$$

where h_n is independent of x and y, and $f_n(x)$ and $g_n(x)$ are different functions, we adopt the terminology used by Rainville [1; p. 170, (3)] and call $H(x, y, t)$ a *bilateral* generating function. We give the following example of a bilateral generating function:

$$(1-t)^{-1+c-\alpha}(1-t+yt)^{-c}\exp\left\{\frac{-xt}{1-t}\right\}{}_1F_1\left[\begin{matrix} c\,; \\ 1+\alpha; \end{matrix}\ \frac{xyt}{(1-t)(1-t+yt)}\right]$$
$$=\sum_{n=0}^{\infty} {}_2F_1\left[\begin{matrix}-n, & c; \\ 1+\alpha & ; \end{matrix}\ y\right] L_n^{(\alpha)}(x)\, t^n. \tag{1}$$

For the set of hypergeometric functions ${}_2F_1\left[\begin{matrix}-n, & c; \\ 1+\alpha & ; \end{matrix}\ y\right]$. Weisner [1; p. 1037, (4.6) with $\gamma=1+\alpha$] obtained this bilateral generating function by the group theoretic method which he developed. For the Laguerre polynomials Brafman [1; p. 180, (5)] used contour integration to obtain (1). Rainville [1; pp. 212–213] obtained (1) by the series manipulation method which is to be explained in this section.

In order to find a bilateral generating function for some set $\{f_n(x)\}$ by the method of this section we need a generating function of the following type:

$$K(x, t, k)=\sum_{n=0}^{\infty} a(n, k)\, f_{n+k}(x)\, t^n.$$

We also need another generating function

$$H(x, t)=\sum_{k=0}^{\infty} b_k f_k(x)\, t^k$$

such that $H(x, t) \neq K(x, t, 0)$. By appropriate substitutions and multiplications we can transform $H(x, t)$ into $J(x, t, y)$ where

$$J(x, t, y)=\sum_{k=0}^{\infty} c_k (ty)^k K(x, t, k)$$
$$=\sum_{k=0}^{\infty} c_k (ty)^k \sum_{n=0}^{\infty} a(n, k)\, f_{n+k}(x)\, t^n.$$

After some simplification we then have

$$J(x, t, y)=\sum_{n=0}^{\infty} d_n f_n(x)\, g_n(y)\, t^n.$$

It is not inherent in the method that $f_n(x)$ and $g_n(x)$ be different functions. In other words, we may obtain a bilinear generating function instead of a bilateral one.

Suppose that for the simple Laguerre polynomials $\{L_n(x)\}$ we have given the generating relation

$$(1-t)^{-1-k}\exp\left\{\frac{-xt}{1-t}\right\}L_k\left(\frac{x}{1-t}\right)=\sum_{n=0}^{\infty}\frac{(n+k)!\,L_{n+k}(x)\,t^n}{k!\,n!}. \tag{2}$$

By means of (2) we hope to transform some other known generating relation into a new one of the form

$$H(x,y,t)=\sum_{n=0}^{\infty}h_n\,g_n(y)\,L_n(x)\,t^n.$$

Let us use the generating relation obtained in Section 3:

$$e^t\,{}_0F_1[-;1;-xt]=\sum_{k=0}^{\infty}\frac{L_k(x)}{k!}t^k, \tag{3}$$

where we have summed over k for convenience in a later substitution. We propose to change the form of (3) so that in its right member will appear the generating function of (2). To introduce the variable y and to provide the $(1-t)^{-k}$ factor, we replace t by $\frac{yt}{1-t}$. We obviously need to replace x by $\frac{x}{1-t}$. Then (3) becomes

$$\exp\left\{\frac{yt}{1-t}\right\}{}_0F_1\left[\begin{matrix}-; & \dfrac{-xyt}{(1-t)^2}\\ 1; & \end{matrix}\right]=\sum_{k=0}^{\infty}\frac{L_k\left(\dfrac{x}{1-t}\right)}{k!}\left(\frac{yt}{1-t}\right)^k. \tag{4}$$

If we compare (4) with (2), we see that we still lack the factors $(1-t)^{-1}$ and $\exp\left\{\frac{-xt}{1-t}\right\}$. Since these factors are independent of k, we multiply both members of (4) by $(1-t)^{-1}\exp\left\{\frac{-xt}{1-t}\right\}$ to get

$$\begin{aligned}(1-t)^{-1}&\exp\left\{\frac{-xt}{1-t}\right\}\exp\left\{\frac{yt}{1-t}\right\}{}_0F_1\left[\begin{matrix}-; & \dfrac{-xyt}{(1-t)^2}\\ 1; & \end{matrix}\right]\\ &=\sum_{k=0}^{\infty}\left[(1-t)^{-1-k}\exp\left\{\frac{-xt}{1-t}\right\}L_k\left(\frac{x}{1-t}\right)\right]\frac{(yt)^k}{k!}.\end{aligned} \tag{5}$$

By substituting (2) in (5), we have

$$\begin{aligned}(1-t)^{-1}&\exp\left\{\frac{-xt+yt}{1-t}\right\}{}_0F_1\left[\begin{matrix}-; & \dfrac{-xyt}{(1-t)^2}\\ 1; & \end{matrix}\right]\\ &=\sum_{k=0}^{\infty}\sum_{n=0}^{\infty}\frac{(n+k)!\,L_{n+k}(x)\,t^n}{k!\,n!}\,\frac{y^k t^k}{k!}.\end{aligned} \tag{6}$$

If in (6) we replace y by $-y$ and simplify, we get

$$(1-t)^{-1}\exp\left\{\frac{-t(x+y)}{(1-t)^2}\right\}{}_0F_1\left[\begin{matrix}-; & \dfrac{xyt}{(1-t)^2}\\ 1; & \end{matrix}\right]=\sum_{n=0}^{\infty}L_n(y)L_n(x)t^n, \tag{7}$$

a *bilinear* generating function which is a special case of the bilinear generating function which we obtained in Section 5 by a different method. If we use exactly the same technique to transform

$$(1-t)^{-c}{}_1F_1\left[\begin{matrix}c; & \dfrac{-xt}{1-t}\\ 1; & \end{matrix}\right]=\sum_{k=0}^{\infty}\frac{(c)_k L_k(x)t^k}{k!},$$

we obtain the bilateral generating relation

$$(1-t)^{-1+c}(1-t+yt)^{-c}\exp\left\{\frac{-xt}{1-t}\right\}{}_1F_1\left[\begin{matrix}c; & \dfrac{xyt}{(1-t)(1-t+yt)}\\ 1; & \end{matrix}\right]$$
$$=\sum_{n=0}^{\infty}{}_2F_1\left[\begin{matrix}-n, & c; & y\\ & 1; & \end{matrix}\right]L_n(x)t^n.$$

See Rainville [1; pp. 212–213 with $\alpha=0$]. This illustration indicates that the given generating function determines whether the transformed generating relation is bilinear or bilateral.

We will now transform a generating function of $\{f_n^{\beta}(x)\}$ into a bilateral generating function. We have from Section 3

$$(1-xt)^{-c}{}_2F_0\left[\begin{matrix}c, & \beta; & \dfrac{t}{1-xt}\\ & -; & \end{matrix}\right]\cong\sum_{k=0}^{\infty}(c)_k f_k^{\beta}(x)t^k, \tag{8}$$

and from Section 4

$$e^{xt}(1-t)^{-\beta-k}f_k^{\beta}(x(1-t))=\sum_{n=0}^{\infty}\frac{(n+k)!\,f_{k+n}^{\beta}(x)t^n}{k!\,n!}. \tag{9}$$

Our purpose is to change the generating relation (8) to a form that will contain in its right member the left member of (9). Accordingly, we must replace x by $x(1-t)$. In order to take care of the factor $(1-t)^{-k}$ and to introduce a third variable we replace t by $\dfrac{yt}{1-t}$. The factors of the left member of (9) that are independent of k can be supplied by multiplying both members by $e^{xt}(1-t)^{-\beta}$. If in (8) we replace x by $x(1-t)$ and t by $\dfrac{yt}{1-t}$, we get

$$\begin{aligned}&(1-xyt)^{-c}{}_2F_0\left[\begin{matrix}c, & \beta; & \dfrac{yt}{(1-t)(1-xyt)}\\ & -; & \end{matrix}\right]\\ &\qquad\cong\sum_{k=0}^{\infty}(c)_k f_k^{\beta}(x(1-t))\left(\frac{yt}{1-t}\right)^k.\end{aligned} \tag{10}$$

If we multiply both members of (10) by $e^{xt}(1-t)^{-\beta}$, we have

$$\begin{aligned} &e^{xt}(1-t)^{-\beta}(1-xyt)^{-c}\,{}_2F_0\left[\begin{matrix} c, & \beta; \\ & -; \end{matrix}\ \frac{yt}{(1-t)(1-xyt)}\right] \\ &\qquad \cong \sum_{k=0}^{\infty}(c)_k y^k t^k \left[e^{xt}(1-t)^{-\beta-k} f_k^{\beta}(x(1-t))\right]. \end{aligned} \tag{11}$$

By means of (9) then (11) becomes

$$\begin{aligned} &e^{xt}(1-t)^{-\beta}(1-xyt)^{-c}\,{}_2F_0\left[\begin{matrix} c, & \beta; \\ & -; \end{matrix}\ \frac{yt}{(1-t)(1-xyt)}\right] \\ &\qquad \cong \sum_{k=0}^{\infty}(c)_k y^k t^k \sum_{n=0}^{\infty}\frac{(n+k)!\, f_{k+n}^{\beta}(x)\, t^n}{k!\, n!}. \end{aligned} \tag{12}$$

After simplifying and replacing y by $-y$, we have the following family of (divergent) bilateral generating relations:

$$\begin{aligned} &e^{xt}(1-t)^{-\beta}(1+xyt)^{-c}\,{}_2F_0\left[\begin{matrix} c, & \beta; \\ & -; \end{matrix}\ \frac{-yt}{(1-t)(1+xyt)}\right] \\ &\qquad \cong \sum_{n=0}^{\infty}{}_2F_0\left[\begin{matrix} -n, & c; \\ & -; \end{matrix}\ y\right] f_n^{\beta}(x)\, t^n. \end{aligned} \tag{13}$$

7. Summary of results. We now list the generating relations for $\{f_n^{\beta}(x)\}$ obtained by the series manipulation methods of this chapter:

$$\sum_{n=0}^{\infty} f_n^{\beta}(x)\, t^n = e^{xt}(1-t)^{-\beta}, \tag{1}$$

$$\sum_{n=0}^{\infty}(c)_n f_n^{\beta}(x)\, t^n \cong (1-xt)^{-c}\,{}_2F_0\left[\begin{matrix} c, & \beta; \\ & -; \end{matrix}\ \frac{t}{1-xt}\right], \tag{2}$$

$$\sum_{n=0}^{\infty}\frac{(n+k)!\, f_{k+n}^{\beta}(x)\, t^n}{k!\, n!} = e^{xt}(1-t)^{-\beta-k} f_k^{\beta}(x(1-t)), \tag{3}$$

$$\begin{aligned} &\sum_{n=0}^{\infty} n!\, f_n^{\beta}(x)\, f_n^{\beta}(y)\, t^n \\ &\qquad \cong e^{xyt}(1-xt)^{-\beta}(1-yt)^{-\beta}\,{}_2F_0\left[\begin{matrix} \beta, & \beta; \\ & -; \end{matrix}\ \frac{t}{(1-xt)(1-yt)}\right], \end{aligned} \tag{4}$$

$$\begin{aligned} &\sum_{n=0}^{\infty}{}_2F_0\left[\begin{matrix} -n, & c; \\ & -; \end{matrix}\ y\right] f_n^{\beta}(x)\, t^n \\ &\qquad \cong e^{xt}(1-t)^{-\beta}(1+xyt)^{-c}\,{}_2F_0\left[\begin{matrix} c, & \beta; \\ & -; \end{matrix}\ \frac{-yt}{(1-t)(1+xyt)}\right]. \end{aligned} \tag{5}$$

The first of these five generating functions is well known. See Boas and Buck [1; pp. 16 and 31], Erdélyi [2; p. 189, (9)], Hochstadt [1; p. 12], Magnus and Oberhettinger [1; p. 85].

Although the generating relation (5) is in bilateral form we can easily change it to bilinear form. Since

$$f_n^c(y)=(-1)^n L_n^{-c-n}(y)=\frac{y^n}{n!}\,{}_2F_0\left(-n, c; -; -\frac{1}{y}\right),$$

we may replace y by $-\dfrac{1}{y}$ and t by yt to get

$$\sum_{n=0}^{\infty} n!\, f_n^c(y)\, f_n^\beta(x)\, t^n \cong e^{xyt}(1-yt)^{-\beta}(1-xt)^{-c}\,{}_2F_0\left(c, \beta; -; \frac{t}{(1-xt)(1-yt)}\right). \tag{5a}$$

Therefore, the generating relation (4) is a special case of (5a) with $c=\beta$.

The generating function of (5a) is equivalent to the generating function given by Meixner [1; p. 533, (14)] for the Charlier polynomials which he denoted by $\{Q_n(x, a)\}$. In harmony with his definition we have the following relation:

$$\begin{aligned} Q_n(a, x) &= (-1)^n x^n\,{}_2F_0\left(-n, -a; -; -\frac{1}{x}\right) \\ &= (-1)^n n!\, f_n^{-a}(x), \end{aligned}$$

where $x>0$ and $a=0, 1, 2, \ldots$.

We showed in Section 2 how a series representation can be obtained from a generating function, and in Section 3 how a generating function can be obtained from a series representation. The method of Section 3 is basic. Of particular importance is the method for obtaining a family of generating functions by introducing a parameter.

All the other methods of the chapter require that we already know at least one generating function. For example, in Section 4 from a given generating relation for some set $\{f_n(x)\}$,

$$G(x, t)=\sum_{n=0}^{\infty} a_n f_n(x)\, t^n$$

we obtained a generalization,

$$K(x, t, k)=\sum_{n=0}^{\infty} b(n, k)\, f_{n+k}(x)\, t^n,$$

such that $K(x,t,0)=G(x,t)$ and $b(n,0)=a_n$. We will call a generating function of this type a k-type generating function.

The method in Section 5 used for finding a bilinear generating function required that for each set $\{f_n(x)\}$ we be given the power series representation of $f_n(x)$ and a k-type generating function. Furthermore, in order to simplify our results we needed to be able to find another generating function by the method of Section 3.

In Section 6 we were able to find a bilateral generating function provided we were given a k-type generating function $K(x,t,k)$ and another generating function $H(x,t)$ such that $K(x,t,0) \neq H(x,t)$.

All evidence available to the author indicates that Rainville should be credited with developing a method for obtaining a k-type generating function (Section 4) and a method for using such a function to determine bilinear generating functions (Section 5). Bedient [1], one of Rainville's students, introduced the method of Section 6 for determining bilateral generating functions.

Chapter II

The Weisner Method

1. Introduction. Weisner [1] has devised a method for obtaining generating functions for sets of functions which satisfy certain conditions. Among the functions which do satisfy these conditions are the Hermite (see Weisner [2]), the Bessel (see Weisner [3]), the generalized Laguerre, and the Gegenbauer.

From the ordinary differential equation which is satisfied by the set of functions under consideration a partial differential equation is constructed. The method is based on finding a nontrivial continuous group of transformations under which this partial differential equation is invariant.

2. The differential equation. To illustrate Weisner's method we shall obtain generating functions for the Laguerre polynomials $\{L_n^{(\alpha)}(x)\}$. These polynomials satisfy the following differential recurrence relations, where the subscripts are nonnegative integers:

$$\frac{d}{dx} L_n^{(\alpha)}(x) = \frac{1}{x}\left[n L_n^{(\alpha)}(x) - (\alpha+n) L_{n-1}^{(\alpha)}(x)\right] \tag{1}$$

and

$$\frac{d}{dx} L_n^{(\alpha)}(x) = \frac{1}{x}\left[(x-\alpha-n-1) L_n^{(\alpha)}(x) + (n+1) L_{n+1}^{(\alpha)}(x)\right]. \tag{2}$$

These two independent differential recurrence relations determine the linear ordinary differential equation

$$x D^2 L_n^{(\alpha)}(x) + (1+\alpha-x) D L_n^{(\alpha)}(x) + n L_n^{(\alpha)}(x) = 0, \tag{3}$$

where $D = \frac{d}{dx}$. If we use operator functional notation and let

$$L\left(x, \frac{d}{dx}, n\right) = x D^2 + (1+\alpha-x) D + n,$$

we may rewrite (3) in the following abbreviated form:

$$L\left(x, \frac{d}{dx}, n\right) L_n^{(\alpha)}(x) = 0. \tag{4}$$

We seek a generating function $G(x, y)$ which by the definition of a generating function must be expressible as follows:

$$G(x, y) = \sum_n g_n L_n^{(\alpha)}(x)\, y^n,$$

where g_n is independent of x and y, but may depend on the parameter α. Any partial differential operator which annuls $G(x, y)$ and for which termwise operation is permissible must also annul $L_n^{(\alpha)}(x)\, y^n$. Since

$$L\left(x, \frac{\partial}{\partial x}, y\frac{\partial}{\partial y}\right) [L_n^{(\alpha)}(x)\, y^n] = y^n L\left(x, \frac{d}{dx}, n\right) [L_n^{(\alpha)}(x)],$$

for $y \neq 0$ we have

$$L\left(x, \frac{\partial}{\partial x}, y\frac{\partial}{\partial y}\right) [L_n^{(\alpha)}(x)\, y^n] = 0$$

if and only if

$$L\left(x, \frac{d}{dx}, n\right) [L_n^{(\alpha)}(x)] = 0.$$

In accordance with the discussion in the preceding paragraph, we now construct from (3) the following partial differential equation by replacing $\frac{d}{dx}$ by $\frac{\partial}{\partial x}$, n by $y\frac{\partial}{\partial y}$ and $L_n^{(\alpha)}(x)$ by $u(x, y)$:

$$\left[x\frac{\partial^2}{\partial x^2} + (1+\alpha-x)\frac{\partial}{\partial x} + y\frac{\partial}{\partial y}\right] u(x, y) = 0. \tag{5}$$

Again using operator functional notation we express (5) in the following abbreviated form:

$$L\left(x, \frac{\partial}{\partial x}, y\frac{\partial}{\partial y}\right) [u(x, y)] = 0. \tag{6}$$

We know, of course, that $u_1(x, y) \equiv L_n^{(\alpha)}(x)\, y^n$ is a solution of (6) since $L_n^{(\alpha)}(x)$ is a solution of (4). For simplicity of notation we will now write L instead of

$$L\left(x, \frac{\partial}{\partial x}, y\frac{\partial}{\partial y}\right) \equiv x\frac{\partial^2}{\partial x^2} + (1+\alpha-x)\frac{\partial}{\partial x} + y\frac{\partial}{\partial y}.$$

3. Linear differential operators. Next we seek linear differential operators which will commute with L or with $\psi(x)L$, where $\psi(x)$ is some function of x to be determined. Suppose C is such an operator. Let

$$C=C_1(x,y)\frac{\partial}{\partial x}+C_2(x,y)\frac{\partial}{\partial y}+C_0(x,y),$$

where each C_i $(i=0,1,2)$ is a function of x and y which is independent of n, but not necessarily independent of the parameter α.

Every linear differential operator of the first order generates a one-parameter Lie group. If we know the linear differential operator C, we will be able to determine (by a method discussed in detail later) the extended form of the group generated by C. In other words, for an arbitrary constant c and an arbitrary function $f(x,y)$ we will be able to determine the effect of the operator e^{cC} on $f(x,y)$, where

$$e^{cC}\equiv 1+cC+\frac{(cC)^2}{2!}+\cdots+\frac{(cC)^k}{k!}+\cdots\equiv\sum_{k=0}^{\infty}\frac{(cC)^k}{k!}.$$

Since C is assumed to be an operator which commutes with $\psi(x)L$, we conclude from

$$e^{cC}\,\psi(x)\,L[u(x,y)]=0$$

that

$$\psi(x)\,L[e^{cC}\,u(x,y)]=0.$$

Then, if $e^{cC}\,u(x,y)$ can be expanded in powers of y, i.e.,

$$e^{cC}\,u(x,y)=\sum_n g_n\,v_n(x)\,y^n,$$

where g_n is independent of x and y, both members of this equation will be annulled by L. Then $v_n(x)$ is a solution of (3). Therefore, $e^{cC}\,u(x,y)$ will generate some of the solutions of (3).

If we wish to determine in advance the existence of a group of operators under which the partial differential equation is invariant, we refer to Lie [1; vol. 3, pp. 492–523]. He determined all partial differential equations in two variables, of the second order, which are invariant under a Lie group. From a practical point of view, for a given partial differential equation it usually takes less time to find the Lie group of operators than to prove that the group exists. However, if we should fail to find a group of operators under which a given partial differential equation is invariant, it would be advantageous to use Lie's results to prove that such a group does not exist.

For the Laguerre polynomials we will now actually find first order linear differential operators B and C such that

$$B[L_n^{(\alpha)}(x)\,y^n]=b_n\,L_{n-1}^{(\alpha)}(x)\,y^{n-1},\qquad n\geqq 1,\tag{7}$$

and

$$C[L_n^{(\alpha)}(x)\,y^n]=c_n\,L_{n+1}^{(\alpha)}(x)\,y^{n+1},\qquad n\geqq 1,\tag{8}$$

where b_n and c_n are functions of n which are independent of x and y, but not necessarily independent of α.

Let

$$B = B_1(x, y)\frac{\partial}{\partial x} + B_2(x, y)\frac{\partial}{\partial y} + B_0(x, y),$$

where each B_i $(i=0, 1, 2)$ is a function of x and y which is independent of n, but not necessarily independent of α. With the aid of (1) we have

$$\begin{aligned} B[L_n^{(\alpha)}(x)\, y^n] = {} & B_1\, x^{-1}\, y^n [n\, L_n^{(\alpha)}(x) - (\alpha+n)\, L_{n-1}^{(\alpha)}(x)] \\ & + B_2\, L_n^{(\alpha)}(x)[n\, y^{n-1}] + B_0\, L_n^{(\alpha)}(x)\, y^n. \end{aligned} \tag{9}$$

In order to make the coefficient of $L_{n-1}^{(\alpha)}(x)\, y^{n-1}$ independent of x and y we choose $B_1 = x\, y^{-1}$. Then (9) becomes

$$\begin{aligned} B[L_n^{(\alpha)}(x)\, y^n] = {} & -(\alpha+n)\, L_{n-1}^{(\alpha)}(x)\, y^{n-1} \\ & + L_n^{(\alpha)}(x)\, y^n [y^{-1}(n + B_2\, n) + B_0]. \end{aligned} \tag{10}$$

In (10) we can make the coefficient of $L_n^{(\alpha)}(x)\, y^n$ equal to zero by choosing $B_2 = -1$ and $B_0 = 0$. Therefore,

$$B = x\, y^{-1}\frac{\partial}{\partial x} - \frac{\partial}{\partial y} \quad \text{and} \quad B[L_n^{(\alpha)}(x)\, y^n] = -(\alpha+n)\, L_{n-1}^{(\alpha)}(x)\, y^{n-1}.$$

Similarly, with the aid of (2) we find

$$\begin{aligned} C[L_n^{(\alpha)}(x)\, y^n] = {} & C_1\, y^n\, x^{-1}[(n+1)\, L_{n+1}^{(\alpha)}(x) + (x-\alpha-n-1)\, L_n^{(\alpha)}(x)] \\ & + C_2\, L_n^{(\alpha)}(x)[n\, y^{n-1}] + C_0\, L_n^{(\alpha)}(x)\, y^n. \end{aligned} \tag{11}$$

By choosing $C_1 = x\, y$ we are able to make the coefficient of $L_{n+1}^{(\alpha)}(x)\, y^{n+1}$ in (11) a function of n only. Then we have

$$\begin{aligned} C[L_n^{(\alpha)}(x)\, y^n] = {} & (n+1)\, L_{n+1}^{(\alpha)}(x)\, y^{n+1} \\ & + L_n^{(\alpha)}(x)\, y^n [(x-\alpha-n-1)\, y + C_2\, n\, y^{-1} + C_0]. \end{aligned} \tag{12}$$

Since we require that C_2 and C_0 be functions of x and y which are independent of n, we choose $C_2 = y^2$ and $C_0 = (-x+\alpha+1)\, y$ in order to make the coefficient of $L_n^{(\alpha)}(x)\, y^n$ equal to zero. Then

$$C = x\, y\frac{\partial}{\partial x} + y^2\frac{\partial}{\partial y} + (-x+\alpha+1)\, y$$

and

$$C[L_n^{(\alpha)}(x)\, y^n] = (n+1)\, L_{n+1}^{(\alpha)}(x)\, y^{n+1}.$$

4. Group of operators. Let $A = y\frac{\partial}{\partial y}$. We will use the commutator notation $[A, B]$ with

$$[A, B]\, u = (AB - BA)\, u.$$

We find that

$$AB\,u = y\frac{\partial}{\partial y}\left[x\,y^{-1}\frac{\partial}{\partial x} - \frac{\partial}{\partial y}\right] u = -x\,y^{-1}\frac{\partial u}{\partial x} + x\frac{\partial^2 u}{\partial y\,\partial x} - y\frac{\partial^2 u}{\partial y^2}$$

and

$$BA\,u = \left[x\,y^{-1}\frac{\partial}{\partial x} - \frac{\partial}{\partial y}\right] y\frac{\partial}{\partial y}\, u = -\frac{\partial u}{\partial y} + x\frac{\partial^2 u}{\partial x\,\partial y} - y\frac{\partial^2 u}{\partial y^2}.$$

Then for $\frac{\partial^2 u}{\partial y\,\partial x} = \frac{\partial^2 u}{\partial x\,\partial y}$, we have

$$[A, B]\, u = -x\,y^{-1}\frac{\partial u}{\partial x} + \frac{\partial u}{\partial y} = -B\,u.$$

In a similar manner we find $[A, C] = C$ and $[B, C] = -2A - (\alpha + 1)$. These commutator relations show that the operators 1, A, B, C generate a Lie group. See Cohn [1].

We would like to prove that these operators commute with L or with $\psi(x)\,L$, where $\psi(x)$ is a function of x yet to be determined. To facilitate this proof we seek to express L or $\psi(x)\,L$ in terms of these operators. We know that

$$L\,u = x\frac{\partial^2 u}{\partial x^2} - (x - \alpha - 1)\frac{\partial u}{\partial x} + y\frac{\partial u}{\partial y}.$$

Also in making the computation of $[B, C]$, we found that

$$CB\,u = x^2\frac{\partial^2 u}{\partial x^2} - y^2\frac{\partial^2 u}{\partial y^2} + (x - \alpha - 1)\left(-x\frac{\partial u}{\partial x} + y\frac{\partial u}{\partial y}\right).$$

Hence we take $\psi(x) = x$ and get $[x\,L - CB]\, u = y^2\frac{\partial^2 u}{\partial y^2} + (\alpha + 1)\, y\frac{\partial u}{\partial y}$. But

$$A^2\, u = y\frac{\partial}{\partial y}\left(y\frac{\partial}{\partial y}\right) u = y^2\frac{\partial^2 u}{\partial y^2} + y\frac{\partial u}{\partial y}.$$

Therefore,

$$(x\,L - CB)\, u = (A^2 + \alpha\, A)\, u,$$

or equivalently,

$$x\,L\,u = (CB + A^2 + \alpha A)\, u.$$

We will now prove that C commutes with xL by using the following commutator relations which have already been established:

$$[A, B] = AB - BA = -B,$$
$$[A, C] = AC - CA = C,$$
$$[B, C] = BC - CB = -2A - (\alpha + 1).$$

Since $xL = CB + A^2 + \alpha A$, we may write

$$\begin{aligned}[C(xL) - (xL)C]u &= [C(CB + A^2 + \alpha A) - (CB + A^2 + \alpha A)C]u \\ &= [C(CB - BC) + (CA^2 - A^2 C) + \alpha(CA - AC)]u.\end{aligned}$$

We first simplify as follows:

$$\begin{aligned}CA^2 - A^2 C &= (CA)A - A(AC) \\ &= (AC - C)A - A(CA + C) = -2CA - C.\end{aligned}$$

Hence, with the further aid of the commutator relations, we now have

$$[C(xL) - (xL)C]u = [C(2A + \alpha + 1) - (2CA + C) + \alpha(-C)]u = [0]u.$$

Therefore, C commutes with xL. In a similar manner we may establish that the operator B commutes with xL.

5. The extended form of the group generated by B and C. We now seek $e^{bB} f(x, y)$ and $e^{cC} f(x, y)$, where $f(x, y)$ is an arbitrary function and b and c are arbitrary constants. The problem of finding $e^{cC} f(x, y)$ is typical; so we give the steps in outline and then in detail. We will use a method suggested by Weisner which is based on Taylor's series in the form

$$\exp\left\{c \frac{d}{dX}\right\} F(X) = F(X + c).$$

In order to use this method we will change the form of the differential operator from

$$C = C_1 \frac{\partial}{\partial x} + C_2 \frac{\partial}{\partial y} + C_0$$

to

$$E = C_1 \frac{\partial}{\partial x} + C_2 \frac{\partial}{\partial y},$$

and finally by change of variable to

$$D = \frac{\partial}{\partial X}.$$

If $\phi(x, y)$ is any solution of

$$C[\phi(x, y)]=0,$$

then $\phi^{-1} C \phi = C_1 \frac{\partial}{\partial X} + C_2 \frac{\partial}{\partial y}$. Let us establish the validity of this statement:

$$\phi^{-1} C \phi u = \phi^{-1} C_1 \frac{\partial}{\partial x}(\phi u) + \phi^{-1} C_2 \frac{\partial}{\partial y}(\phi u) + \phi^{-1} C_0(\phi u)$$

$$= \phi^{-1} C_1 \left(u \frac{\partial \phi}{\partial x} + \phi \frac{\partial u}{\partial x}\right) + \phi^{-1} C_2 \left(u \frac{\partial \phi}{\partial y} + \phi \frac{\partial u}{\partial y}\right) + \phi^{-1} C_0 \phi u$$

$$= \phi^{-1} C_1 \phi \frac{\partial u}{\partial x} + \phi^{-1} C_2 \phi \frac{\partial u}{\partial y} + \phi^{-1} u \left(C_1 \frac{\partial \phi}{\partial x} + C_2 \frac{\partial \phi}{\partial y} + C_0 \phi\right).$$

Now if $C[\phi(x, y)]=0$, we have

$$C_1 \frac{\partial \phi}{\partial x} + C_2 \frac{\partial \phi}{\partial y} + C_0 \phi = 0.$$

Therefore, if ϕ is annulled by C, we have

$$\phi^{-1} C \phi u = \phi^{-1} C_1 \phi \frac{\partial u}{\partial x} + \phi^{-1} C_2 \phi \frac{\partial u}{\partial y}$$

$$= C_1 \frac{\partial u}{\partial x} + C_2 \frac{\partial u}{\partial y}.$$

If we let $E = \phi^{-1} C \phi$, then $C = \phi E \phi^{-1}$.

By substituting for the operator C and simplifying, we get

$$e^{cC} f(x, y) = e^{c\phi E \phi^{-1}} f(x, y)$$
$$= \phi(x, y)\, e^{cE}[\phi^{-1}(x, y) f(x, y)].$$

We next choose new variables X and Y so that the operator E is transformed into the operator $D \equiv \partial/\partial x$. Under this change of variable let $\phi^{-1}(x, y) f(x, y)$ be transformed into $F(X, Y)$. By means of Taylor's theorem, we then have

$$e^{cE}[\phi^{-1}(x, y) f(x, y)] = e^{cD} F(X, Y)$$
$$= F(X + c, Y).$$

Let $F(X + c, Y)$ be transformed into $g(x, y)$ by the inverse substitution. Then

$$e^{cC} f(x, y) = \phi(x, y)\, g(x, y).$$

By the method outlined we will now actually compute $e^{cC} f(x, y)$, where $C = x y \frac{\partial}{\partial x} + y^2 \frac{\partial}{\partial y} + (\alpha + 1 - x) y$. First, we seek a function $\phi(x, y)$ such that $C\phi = 0$, i.e.,

$$x y \frac{\partial \phi}{\partial x} + y^2 \frac{\partial \phi}{\partial y} + (\alpha + 1 - x) y \phi = 0.$$

For $y \neq 0$, we get

$$x \frac{\partial \phi}{\partial x} + y \frac{\partial \phi}{\partial y} + (\alpha + 1 - x) \phi = 0.$$

In order to find ϕ we solve the system:

$$\frac{dx}{x} = \frac{dy}{y} = \frac{d\phi}{(x - \alpha - 1)\phi}.$$

(See Frederic H. Miller [1], p. 85.) We find the general integral to be

$$\sigma\left(x y^{-1}, \frac{e^x}{x^{\alpha+1}\phi}\right) = 0,$$

where σ is an arbitrary function. We choose the special case

$$(x y^{-1}) \frac{e^x}{x^{\alpha+1}\phi} - 1 = 0.$$

Then $\phi = x^{-\alpha} y^{-1} e^x$, $\phi^{-1} = x^\alpha y e^{-x}$, and $E = \phi^{-1} C \phi = x y \frac{\partial}{\partial x} + y^2 \frac{\partial}{\partial y}$.

Next we choose new variables X and Y so that E will be transformed into $k \frac{\partial}{\partial X}$, where $k^2 = 1$. If we substitute

$$\frac{\partial u}{\partial x} = \frac{\partial u}{\partial X} \frac{\partial X}{\partial x} + \frac{\partial u}{\partial Y} \frac{\partial Y}{\partial x}$$

and

$$\frac{\partial u}{\partial y} = \frac{\partial u}{\partial X} \frac{\partial X}{\partial y} + \frac{\partial u}{\partial Y} \frac{\partial Y}{\partial y}$$

in the relation

$$E u = x y \frac{\partial u}{\partial x} + y^2 \frac{\partial u}{\partial y},$$

we get

$$E u = \left[x y \frac{\partial X}{\partial x} + y^2 \frac{\partial X}{\partial y}\right] \frac{\partial u}{\partial X} + \left[x y \frac{\partial Y}{\partial x} + y^2 \frac{\partial Y}{\partial y}\right] \frac{\partial u}{\partial Y}.$$

By setting the coefficient of $\partial u / \partial Y$ equal to zero, we get

$$x \frac{\partial Y}{\partial x} + y \frac{\partial Y}{\partial y} = 0.$$

The corresponding subsidiary equations are $dx/x=dy/y$; $dY=0$. (See Frederic H. Miller, [1], p. 96.) We find $Y=\xi(x/y)$, where ξ is an arbitrary function. We choose the special case $Y=x\,y^{-1}$.

By setting the coefficient of $\partial u/\partial X$ equal to k, we get

$$x\frac{\partial X}{\partial x}+y\frac{\partial X}{\partial y}=y^{-1}k.$$

The corresponding subsidiary equations are

$$\frac{dx}{x}=\frac{dy}{y}=\frac{dX}{k\,y^{-1}},$$

from which we obtain the general integral

$$\eta(x\,y^{-1},\,-y^{-1}-k\,X)=0,$$

where η is an arbitrary function. We choose $k=-1$ and $\eta=-y^{-1}+X$. We now have

$$X=y^{-1} \quad \text{and} \quad Y=x\,y^{-1}.$$

Solving for x and y, we get

$$x=X^{-1}\,Y \quad \text{and} \quad y=X^{-1}.$$

We are now in a position to find $e^{cC}f(x,y)$, the extended form of the transformation group generated by C. Recalling that $C=\phi\,E\,\phi^{-1}$, where $\phi=x^{-\alpha}\,y^{-1}\,e^{x}$, we may write $e^{cC}f(x,y)$ in the following form:

$$e^{cC}f(x,y)=e^{c\phi E\phi^{-1}}f(x,y)=(x^{-\alpha}\,y^{-1}\,e^{x})\,e^{cE}[(x^{\alpha}\,y\,e^{-x})\,f(x,y)].$$

We have shown that the substitution $x=X^{-1}\,Y$ and $y=X^{-1}$ will transform E into $(-1)\,D$, where $D=\partial/\partial X$. Making this substitution and applying Taylor's theorem, we find that

$$\begin{aligned}
&e^{cE}[x^{\alpha}\,y\,e^{-x}f(x,y)]\\
&\quad=e^{-cD}[(X^{-1}\,Y)^{\alpha}(X^{-1})\exp\{-X^{-1}\,Y\}\,f(X^{-1}\,Y,X^{-1})]\\
&\quad=e^{-cD}[X^{-\alpha-1}\,Y^{\alpha}\exp\{-X^{-1}\,Y\}\,f(X^{-1}\,Y,X^{-1})]\\
&\quad=(X-c)^{-\alpha-1}\,Y^{\alpha}\exp\{-(X-c)^{-1}\,Y\}\,f\big((X-c)^{-1}\,Y,(X-c)^{-1}\big).
\end{aligned}$$

Finally, we use the inverse substitution $X=y^{-1}$ and $Y=x\,y^{-1}$ to establish the following result:

$$\begin{aligned}
e^{cC}f(x,y)&=x^{-\alpha}\,y^{-1}\,e^{x}[(y^{-1}-c)^{-\alpha-1}(x\,y^{-1})^{\alpha}\\
&\qquad\cdot\exp\left\{\frac{-x\,y^{-1}}{y^{-1}-c}\right\}f\left(\frac{x\,y^{-1}}{y^{-1}-c},\frac{1}{y^{-1}-c}\right)\\
&=(1-c\,y)^{-\alpha-1}\exp\left\{\frac{-c\,x\,y}{1-c\,y}\right\}f\left(\frac{x}{1-c\,y},\frac{y}{1-c\,y}\right).
\end{aligned}$$

In a similar manner, we find

$$e^{bB} f(x, y) = f\left(\frac{x y}{y-b}, y-b\right).$$

Then

$$e^{cC} e^{bB} f(x, y)$$
$$= (1-c y)^{-1-\alpha} \exp\left\{\frac{-c x y}{1-c y}\right\} f\left(\frac{x y}{(1-c y)(y-b+b c y)}, \frac{y-b+b c y}{1-c y}\right).$$

6. Generating functions. At this point it is expedient to divide the discussion into two parts. The logical basis for this division will become evident in Part II.

Part I. *Generating functions derived from the first order operator* $(A-v)$. From previous considerations we know that $u_1(x, y) \equiv L_v^{(\alpha)}(x) y^v$ is annulled by L and by $A-v$, where $A = y\frac{\partial}{\partial y}$. To obtain generating functions for $\{L_n^{(\alpha)}(x)\}$ we now transform $u_1(x, y)$ by means of the operator $e^{cC} e^{bB}$. We shall consider the following cases:

Case 1. $b=1$, $c=0$,

Case 2. $b=0$, $c=1$,

Case 3. $b c \neq 0$.

Part I, *Case* 1. Since for any arbitrary function $f(x, y)$

$$e^{bB} f(x, y) = f\left(\frac{x y}{y-b}, y-b\right),$$

for $b=1$ we have

$$e^{B}[y^v L_v^{(\alpha)}(x)] = (y-1)^v L_v^{(\alpha)}\left(\frac{x y}{y-1}\right).$$

Let $G(x, y) = (y-1)^v L_v^{(\alpha)}\left(\frac{x y}{y-1}\right)$. $G(x, y)$ may also be written in the following form:

$$G(x, y) = y^v (1-y^{-1})^v L_v^{(\alpha)}\left(\frac{x}{1-y^{-1}}\right).$$

We first require that v be a positive integer and expand $G(x, y)$ as follows:

$$y^v (1-y^{-1})^v L_v^{(\alpha)}\left(\frac{x}{1-y^{-1}}\right) = \sum_{n=0}^{v} g_n L_{v-n}^{(\alpha)}(x) y^{v-n},$$

where g_n may contain the parameters α and v, but is independent of x and y. In order to find g_n we divide by y^v and set $x=0$:

$$(1-y^{-1})^v \frac{(\alpha+1)_v}{v!} = \sum_{n=0}^{v} g_n \frac{(\alpha+1)_{v-n}}{(v-n)!} y^{-n}.$$

By expanding the left member and equating coefficients of y^{-n}, we get

$$g_n = \frac{(-\alpha-\nu)_n}{n!}.$$

Replacing y^{-1} by t, we get the generating relation

$$(1-t)^\nu L_\nu^{(\alpha)}\left(\frac{x}{1-t}\right) = \sum_{n=0}^{\nu} \frac{(-\alpha-\nu)_n}{n!} L_{\nu-n}^{(\alpha)}(x)\, t^n. \tag{1}$$

Truesdell [1; p. 85, (9), with y^n replaced by $(-y)^n$] obtained a similar result by an entirely different method.

$G(x, y)$ may also be expanded in positive powers of y. From $G(x, y)$ we remove a constant factor and write

$$(1-y)^\nu {}_1F_1\left[\begin{matrix}-\nu; & \dfrac{-xy}{1-y}\\ \alpha+1; & \end{matrix}\right] = \sum_{n=0}^{\infty} h_n L_n^{(\alpha)}(x)\, y^n.$$

In this expansion we permit ν to be any real nonzero number. In order to determine h_n, we set $x=0$:

$$(1-y)^\nu = \sum_{n=0}^{\infty} h_n \frac{(\alpha+1)_n}{n!} y^n.$$

By expanding the left member and comparing coefficients of y^n, we find

$$h_n = \frac{(-\nu)_n}{(\alpha+1)_n}.$$

Replacing $-\nu$ by c, we obtain the generating relation

$$(1-y)^{-c} {}_1F_1\left[\begin{matrix}c; & \dfrac{-xy}{1-y}\\ \alpha+1; & \end{matrix}\right] = \sum_{n=0}^{\infty} \frac{(c)_n}{(\alpha+1)_n} L_n^{(\alpha)}(x)\, y^n. \tag{2}$$

See Erdélyi [3; p. 263, (14)] and Rainville [1; p. 135, (13)].

Part I, *Case* 2. $b=0$, $c=1$. For any arbitrary function $f(x, y)$ we have proved that

$$e^{cC} f(x, y) = (1-cy)^{-\alpha-1} \exp\left\{\frac{-cxy}{1-cy}\right\} f\left(\frac{x}{1-cy}, \frac{y}{1-cy}\right).$$

Therefore, for $c=1$, we have

$$e^{C}[y^\nu L_\nu^{(\alpha)}(x)] = (1-y)^{-\alpha-1} \exp\left\{\frac{-xy}{1-y}\right\}\left[L_\nu^{(\alpha)}\left(\frac{x}{1-y}\right)\right]\left(\frac{y}{1-y}\right)^\nu.$$

Also from

$$C[y^\nu L_\nu^{(\alpha)}(x)] = (\nu+1) L_{\nu+1}^{(\alpha)}(x)\, y^{\nu+1},$$

we can prove by induction on n that

$$e^C[y^\nu L_\nu^{(\alpha)}(x)] = \sum_{n=0}^{\infty} \frac{(\nu+1)\cdots(\nu+n)}{n!} L_{\nu+n}^{(\alpha)}(x)\, y^{\nu+n}$$

$$= \sum_{n=0}^{\infty} \frac{(\nu+1)_n}{n!} L_{\nu+n}^{(\alpha)}(x)\, y^{\nu+n}.$$

If we equate the two expressions, each of which is equal to $e^C[y^\nu L_\nu^{(\alpha)}(x)]$, we get

$$(1-y)^{-\alpha-1} \exp\left\{\frac{-xy}{1-y}\right\} \frac{y^\nu}{(1-y)^\nu} L_\nu^{(\alpha)}\left(\frac{x}{1-y}\right) = \sum_{n=0}^{\infty} \frac{(\nu+1)_n}{n!} L_{\nu+n}^{(\alpha)}(x)\, y^{\nu+n}.$$

If we divide both members of this equation by y^ν and simplify, we get

$$(1-y)^{-\nu-\alpha-1} \exp\left\{\frac{-xy}{1-y}\right\} L_\nu^{(\alpha)}\left(\frac{x}{1-y}\right) = \sum_{n=0}^{\infty} \frac{(\nu+1)_n}{n!} L_{\nu+n}^{(\alpha)}(x)\, y^n. \quad (3)$$

See Rainville [1; p. 211, (9)] and Truesdell [1; p. 84, (8)]. With $\nu=0$ in (3) we get the familiar generating function

$$(1-y)^{-\alpha-1} \exp\left\{\frac{-xy}{1-y}\right\} = \sum_{n=0}^{\infty} L_n^{(\alpha)}(x)\, y^n. \quad (3\text{a})$$

Part I, *Case* 3. $bc \neq 0$. In order to simplify results we choose $c=1$ and $b=-\frac{1}{w}$. Then for all finite values of w, we have $bc \neq 0$. For an arbitrary function $f(x,y)$ we then have

$$e^C e^{(-\frac{1}{w})B} f(x,y) = (1-y)^{-\alpha-1} \exp\left\{\frac{-xy}{1-y}\right\} f(\xi,\eta)$$

where $\xi = \dfrac{xwy}{(1-y)(1-y+wy)}$, and $\eta = \dfrac{1-y+wy}{w(1-y)}$. Then, for ν a non-negative integer, we have

$$\begin{aligned} &e^C e^{-\frac{1}{w}B}[y^\nu L_\nu^{(\alpha)}(x)] \\ &= (1-y)^{-\alpha-1} \exp\left\{\frac{-xy}{1-y}\right\} \left(\frac{1-y+wy}{1-y}\right)^\nu L_\nu^{(\alpha)}\left(\frac{xwy}{(1-y)(1-y+wy)}\right) \\ &= (1-y)^{-\alpha-1-\nu} \\ &\quad \cdot \exp\left\{\frac{-xy}{1-y}\right\} (1-y+wy)^\nu \frac{(1+\alpha)_\nu}{\nu!}\, {}_1F_1\left[\begin{matrix} -\nu; \\ 1+\alpha; \end{matrix} \; \frac{xwy}{(1-y)(1-y+wy)}\right]. \end{aligned} \quad (4)$$

Except for the constant factor $\frac{(1+\alpha)_\nu}{\nu!}$, the right member of (4) is equivalent to a bilateral generating function obtained by Brafman [1; p. 180, (5)], by Rainville [1; p. 213], and by Weisner [1; p. 1037, (4.6), with $\gamma=1+\alpha$]:

$$(1-y)^{-1-\nu-\alpha}(1-y+wy)^{\nu}\exp\left\{\frac{-xy}{1-y}\right\}{}_1F_1\left[\begin{matrix}-\nu; \\ 1+\alpha;\end{matrix}\ \frac{xwy}{(1-y)(1-y+wy)}\right] = \sum_{n=0}^{\infty} {}_2F_1\left[\begin{matrix}-n, & -\nu; \\ & 1+\alpha;\end{matrix}\ w\right] L_n^{(\alpha)}(x)\, y^n. \tag{5}$$

Part II. *Generating functions derived from operators not conjugate to* $(A-\nu)$. Let $S=e^{cC}e^{bB}$. Then for each choice of b and c, $S(A-\nu)S^{-1}$ represents an operator conjugate to $(A-\nu)$. Consider the set of operators $\rho=\{R|R=r_1A+r_2B+r_3C+r_4$, for all choices of the ratios of the coefficients except $r_1=r_2=r_3=0\}$. For each R, the conjugate SRS^{-1} is again an element of the set ρ, where S is an element of the Lie group Γ.

We are interested in solutions of the simultaneous partial differential equations $Lu=0$ and $Ru=0$. However, $Lu=0$ and $Ru=0$ if and only if $\psi(x)L(Su)=0$ and $SRS^{-1}(Su)=0$, since S commutes with $\psi(x)L$ and operator multiplication is associative. Therefore, we need to consider only one operator from each of the conjugate subsets into which the set of operators is partitioned by the element S of the Lie group Γ.

Before finding the conjugate sets of operators of the first order for the Laguerre polynomials we introduce a notation used by Morse and Feshbach [1; p. 112, (1.27b)]: For any two linear operators X and Y with a common domain of operands, we have the formal expansion

$$e^{tX}\,Y e^{-tX}=\sum_{k=0}^{\infty}\frac{t^k}{k!}[X,Y]_k,$$

where $[X,Y]_0=Y$, $[X,Y]_1=XY-YX$, and $[X,Y]_k=[X,[X,Y]_{k-1}]$, $(k=2,3,\dots)$. Using this notation we compute $e^{bB}Ce^{-bB}$ as an illustration:

$$\begin{aligned} e^{bB}Ce^{-bB} &= \sum_{k=0}^{\infty}\frac{b^k}{k!}[B,C]_k \\ &= [B,C]_0+b[B,C]_1+\frac{b^2}{2!}[B,C]_2+\cdots. \end{aligned}$$

By definition $[B,C]_0=C$ and $[B,C]_1=BC-CB$. Using the fact that $BC-CB=-2A-\alpha-1$, we get

$$\begin{aligned} [B,C]_2 &= [B,[B,C]_1]=[B,-2A-\alpha-1] \\ &= 2[B,A]=-2B. \end{aligned}$$

Furthermore,

$$[B, C]_3 = [B, [B, C]_2] = [B, -2B] = 0.$$

Hence, if $k \geqq 3$, $[B, C]_k = 0$. We are now able to express $e^{bB} C e^{-bB}$ in the following form:

$$\begin{aligned} e^{bB} C e^{-bB} &= C + b(-2A - \alpha - 1) + \frac{b^2}{2}(-2)B \\ &= -2bA - b^2 B + C - b(\alpha+1). \end{aligned}$$

In a similar manner we compute the other entries in the following summary:

$$\begin{aligned} e^{aA} B e^{-aA} &= e^{-a} B, \\ e^{aA} C e^{-aA} &= e^{a} C, \\ e^{bB} A e^{-bB} &= A + bB, \\ e^{bB} C e^{-bB} &= -2bA - b^2 B + C - b(\alpha+1), \\ e^{cC} A e^{-cC} &= A - cC, \\ e^{cC} B e^{-cC} &= 2cA + B - c^2 C + c(\alpha+1). \end{aligned}$$

Then for $S = e^{cC} e^{bB}$, we have

$$\begin{aligned} SAS^{-1} &= e^{cC} e^{bB} A e^{-bB} e^{-cC} = e^{cC}(A + bB)\, e^{-cC} \\ &= (A - cC) + b[2cA + B - c^2 C + c(\alpha+1)] \\ &= (1 + 2bc)A + bB - c(1 + bc)C + bc(\alpha+1). \end{aligned}$$

Therefore, for

$$R = r_1 A + r_2 B + r_3 C + r_4,$$

$(A - v)$ is conjugate to R if

$$r_1 = 1 + 2bc, \quad r_2 = b, \quad r_3 = -c(1 + bc).$$

From two of these three equations we can find b and c in terms of r_1, r_2, and r_3. The third equation then imposes a restrictive relation on the r_i $(i = 1, 2, 3)$. If $r_2 = 0$, then $b = 0$, $c = -r_3$, and $r_1 = 1$. If $r_2 \neq 0$, then $b = r_2$, $c = \dfrac{r_1 - 1}{2r_2}$, and

$$r_3 = -\left(\frac{r_1 - 1}{2r_2}\right)\left(1 + \frac{r_1 - 1}{2}\right),$$

or equivalently,

$$r_1^2 + 4r_2 r_3 = 1.$$

In other words, for all possible choices of b and c

$$r_1^2 + 4r_2 r_3 \neq 0.$$

Therefore, $(A-v)$ is not conjugate to the set of operators for which $r_1^2+4r_2 r_3=0$.

The operators for which $r_1^2+4r_2 r_3=0$ may be considered under the following three cases:

Case 1. $r_1=0,\ r_2=1,\ r_3=0$.

Case 2. $r_1=2,\ r_2=1,\ r_3=-1$.

Case 3. $r_1=0,\ r_2=0,\ r_3=1$.

Part II, *Case* 1. We seek a solution for the system $Lu=0$ and $(B+\eta)u=0$, where η is a nonzero constant. For convenience we choose $\eta=1$ and write the equations out in full as follows:

$$x\frac{\partial^2 u}{\partial x^2}+(1+\alpha-x)\frac{\partial u}{\partial x}+y\frac{\partial u}{\partial y}=0$$

and

$$x y^{-1}\frac{\partial u}{\partial x}-\frac{\partial u}{\partial y}+u=0.$$

One solution of this system is

$$u(x,y)=e^y\,{}_0F_1(-;1+\alpha;-xy).$$

If this function is expanded in powers of y, we get

$$e^y\,{}_0F_1(-;1+\alpha;-xy)=\sum_{n=0}^{\infty}\frac{L_n^{(\alpha)}(x)}{(1+\alpha)_n}y^n. \tag{6}$$

See Boas and Buck [1; p. 32], Erdélyi [2; p. 189, (18)], and Rainville [1; p. 201, (1) and (2)].

Part II, *Case* 2. We now seek a solution of the system $Lu=0$ and $(2A+B-C+\lambda)u=0$, where λ is a nonzero constant. We avoid the task of actually solving this system by recalling that

$$e^C B e^{-C}=2A+B-C+(\alpha+1),$$

and hence

$$e^C(B-w)e^{-C}=2A+B-C+(\alpha+1-w).$$

Let $S_1=e^C$. Then we have

$$S_1(B-w)S_1^{-1}(S_1 u)=0$$

if $(B-w)u=0$. Therefore, if u is annulled by L and $B-w$, then $S_1 u$ is annulled by L and $2A+B-C+(1+\alpha-w)$.

From Part II, Case 1, with $\eta=-w$, we get that

$$u(x,-wy)\equiv e^{-wy}\,{}_0F_1(-;1+\alpha;wxy)$$

is a solution of the system

$$L u=0 \quad \text{and} \quad (B-w)\,u=0.$$

We know that for an arbitrary function $f(x,y)$

$$\begin{aligned} S_1 f(x,y) &= e^C f(x,y) \\ &= (1-y)^{-\alpha-1} \exp\left\{\frac{-x y}{1-y}\right\} f\left(\frac{x}{1-y}, \frac{y}{1-y}\right). \end{aligned}$$

Then

$$\begin{aligned} & S_1 u(x,-w y) \\ &\quad = (1-y)^{-\alpha-1} \exp\left\{\frac{-x y}{1-y}\right\} \exp\left\{\frac{-w y}{1-y}\right\} {}_0F_1\left[\begin{matrix} -; \\ 1+\alpha; \end{matrix} \; \frac{w x y}{(1-y)^2}\right]. \end{aligned}$$

By noting the symmetry in x and w in this generating function and by using Part I, Case 2, with $v=0$, we obtain the following bilinear generating relation:

$$\begin{aligned} & (1-y)^{-\alpha-1} \exp\left\{\frac{-y(x+w)}{1-y}\right\} {}_0F_1\left[\begin{matrix} -; \\ 1+\alpha; \end{matrix} \; \frac{w x y}{(1-y)^2}\right] \\ &\quad = \sum_{n=0}^{\infty} \frac{n!}{(1+\alpha)_n} L_n^{(\alpha)}(w)\, L_n^{(\alpha)}(x)\, y^n. \end{aligned}$$

See Rainville [1; p. 212, Theorem 69] and Erdélyi [2; p. 189, (20)].

Part II, *Case* 3. We seek a solution of the system

$$L u=0 \quad \text{and} \quad (C+\lambda)\,u=0,$$

where λ is a nonzero constant. Since we already have a solution for the system

$$L u=0 \quad \text{and} \quad (B+\eta)\,u=0,$$

we attempt to find b and c such that

$$e^{bB} e^{cC} B\, e^{-cC} e^{-bB} = k\,C,$$

where k is a nonzero constant. We find that

$$\begin{aligned} & e^{bB} e^{cC} B\, e^{-cC} e^{-bB} \\ &\quad = e^{bB}[2cA+B-c^2 C+c(\alpha+1)]\, e^{-bB} \\ &\quad = 2c[A+bB]+B-c^2[-2bA-b^2 B+C+b(-\alpha-1)]+c(\alpha+1) \\ &\quad = 2c(1+bc)A+(1+bc)^2 B-c^2 C+(\alpha+1)c(bc+1). \end{aligned}$$

If we choose $b=1$, $c=-1$, and let $S_2=e^B e^{-C}$, we then have $S_2 B S_2^{-1}=-C$, and hence

$$S_2(B+\eta)\,S_2^{-1} = -C+\eta.$$

If $(B+\eta)u=0$, then $S_2(B+\eta)S_2^{-1}(S_2 u)=0$. Therefore, if $u(x,y)$ is annulled by L and $B+\eta$, then $S_2 u(x,y)$ is annulled by L and $C-\eta$.

For $\eta=1$, we have from Part II, Case 1,

$$u(x,y)=e^y\,{}_0F_1(-;1+\alpha;-xy).$$

The generating function we now seek is a transformation of $u(x,y)$,

$$S_2 u(x,y)=e^B e^{-C} u(x,y).$$

We begin by determining the effect of the operator e^{-C}:

$$\begin{aligned}S_2 u(x,y)&=e^B e^{-C}[e^y\,{}_0F_1(-;1+\alpha;-xy)]\\&=e^B(1+y)^{-\alpha-1}\exp\left\{\frac{y(x+1)}{1+y}\right\}{}_0F_1\left[\begin{matrix}-; & \dfrac{-xy}{(1+y)^2}\\ 1+\alpha; & \end{matrix}\right].\end{aligned}$$

Continuing the transformation of $u(x,y)$, we get

$$\begin{aligned}S_2 u(x,y)&=y^{-\alpha-1}\exp\left\{\frac{y-1}{y}\left(\frac{xy}{y-1}+1\right)\right\}{}_0F_1\left[\begin{matrix}-; & \dfrac{-(y-1)}{y^2}\cdot\dfrac{xy}{y-1}\\ 1+\alpha; & \end{matrix}\right]\\&=y^{-\alpha-1}\exp\left\{1+x-\frac{1}{y}\right\}{}_0F_1\left[\begin{matrix}-; & \dfrac{-x}{y}\\ 1+\alpha; & \end{matrix}\right].\end{aligned}$$

If we let $y=-\dfrac{1}{t}$ and expand in powers of t, we get

$$\begin{aligned}&(-t)^{\alpha+1}\exp\{x+t\}\,{}_0F_1\left[\begin{matrix}-; & xt\\ 1+\alpha; & \end{matrix}\right]\\&\quad=\sum_{n=0}^{\infty}\frac{(-1)^n}{n!}\,{}_1F_1\left[\begin{matrix}n+\alpha+1; & x\\ \alpha+1; & \end{matrix}\right](-t)^{n+\alpha+1},\end{aligned}$$

or equivalently,

$$e^t\,{}_0F_1(-;1+\alpha;xt)=\sum_{n=0}^{\infty}e^{-x}\,{}_1F_1\left[\begin{matrix}n+\alpha+1; & x\\ \alpha+1; & \end{matrix}\right]\frac{t^n}{n!}.$$

By Kummer's first formula (see Rainville [1; p. 125, Theorem 42]) we have

$$e^{-x}\,{}_1F_1\left[\begin{matrix}n+\alpha+1; & x\\ \alpha+1; & \end{matrix}\right]={}_1F_1\left[\begin{matrix}-n; & -x\\ 1+\alpha; & \end{matrix}\right].$$

Therefore, we have obtained the generating relation

$$\begin{aligned}e^t\,{}_0F_1(-;1+\alpha;xt)&=\sum_{n=0}^{\infty}{}_1F_1\left[\begin{matrix}-n; & -x\\ 1+\alpha; & \end{matrix}\right]\frac{t^n}{n!}\\&=\sum_{n=0}^{\infty}\frac{L_n^{(\alpha)}(-x)}{(1+\alpha)_n}t^n.\end{aligned}$$

But with x replaced by $(-x)$, we have exactly the same generating relation already obtained in Part II, Case 1.

7. Summary. By the Weisner method we have obtained for the Laguerre polynomials the generating functions listed below:

Part I. $r_1^2+4r_2 r_3 \neq 0$.

Case 1.

$$(1-t)^v L_n^{(\alpha)}\left(\frac{x}{1-t}\right)=\sum_{n=0}^{v}\frac{(-\alpha-v)_n}{n!} L_{v-n}^{(\alpha)}(x)\, t^n. \tag{1}$$

$$(1-t)^{-c}\,{}_1F_1\left(c;1+\alpha;\frac{-xt}{1-t}\right)=\sum_{n=0}^{\infty}\frac{(c)_n L_n^{(\alpha)}(x)\, t^n}{(1+\alpha)_n}. \tag{2}$$

Case 2.

$$(1-t)^{-1-\alpha-v}\exp\left\{\frac{-xt}{1-t}\right\} L_v^{(\alpha)}\left(\frac{x}{1-t}\right)=\sum_{n=0}^{\infty}\frac{(v+1)_n}{n!} L_{v+n}^{(\alpha)}(x)\, t^n. \tag{3}$$

Here if $v=0$, we have

$$(1-t)^{-1-\alpha}\exp\left\{\frac{-xt}{1-t}\right\}=\sum_{n=0}^{\infty} L_n^{(\alpha)}(x)\, t^n. \tag{3a}$$

Case 3.

$$\begin{aligned}(1-t)^{-1-v-\alpha}(1-t+wt)^v \exp\left\{\frac{-xt}{1-t}\right\}{}_1F_1\left[\begin{matrix}-v; \\ 1+\alpha;\end{matrix}\ \frac{wxt}{(1-t)(1-t+wt)}\right] \\ =\sum_{n=0}^{\infty}{}_2F_1\left[\begin{matrix}-n, & -v; \\ & 1+\alpha;\end{matrix}\ w\right] L_n^{(\alpha)}(x)\, t^n.\end{aligned} \tag{4}$$

Part II. $r_1^2+4r_2 r_3=0$.

Case 1.

$$e^t\,{}_0F_1(-;1+\alpha;-xt)=\sum_{n=0}^{\infty}\frac{L_n^{(\alpha)}(x)\, t^n}{(1+\alpha)_n}. \tag{5}$$

Case 2.

$$\begin{aligned}(1-t)^{-\alpha-1}\exp\left\{\frac{-(x+w)t}{1-t}\right\}{}_0F_1\left[\begin{matrix}-; \\ 1+\alpha;\end{matrix}\ \frac{wxt}{(1-t)^2}\right] \\ =\sum_{n=0}^{\infty}\frac{n!}{(1+\alpha)_n} L_n^{(\alpha)}(w)\, L_n^{(\alpha)}(x)\, t^n.\end{aligned} \tag{6}$$

In order to demonstrate the power of the Weisner method, we have used a familiar set of polynomials. This method has produced for the Laguerre polynomials the six well known generating functions that have been laboriously found, one at a time by various other methods, since Laguerre's work was published in 1879.

Chapter III

Further Results by the Weisner Method

1. Introduction. We will now use the Weisner method to obtain generating functions for the modified Laguerre polynomials $\{(-1)^n L_n^{-\beta-n}(x)\}$, the simple Bessel polynomials, and the Gegenbauer polynomials. Since the details of this method were given in Chapter II, we will discuss only briefly the procedure followed in obtaining generating functions for the sets of this chapter.

2. The modified Laguerre polynomials $\{(-1)^n L_n^{-\beta-n}(x)\}$. We turn again to the set of modified Laguerre polynomials used in Chapter I to illustrate series manipulation methods. Let

$$\begin{aligned} f_n^\beta(x) &= (-1)^n L_n^{-\beta-n}(x) \\ &= \frac{(\beta)_n}{n!}\, {}_1F_1\left[\begin{matrix} -n; \\ 1-\beta-n; \end{matrix}\; x\right]. \end{aligned}$$

Then $f_n^\beta(x)$ satisfies the two independent differential recurrence relations

$$D f_n^\beta(x) = f_{n-1}^\beta(x) \tag{1}$$

and

$$x\,D f_n^\beta(x) = (x+n+\beta)\, f_n^\beta(x) - (n+1)\, f_{n+1}^\beta(x), \tag{2}$$

where $D = d/dx$. Also (1) and (2) determine the ordinary differential equation

$$x\,D^2 f_n^\beta(x) + (1-\beta-n-x)\, D f_n^\beta(x) + n\, f_n^\beta(x) = 0. \tag{3}$$

In order to use Weisner's method we construct from (3) the following partial differential equation by replacing $\frac{d}{dx}$ by $\frac{\partial}{\partial x}$, n by $y\frac{\partial}{\partial y}$, and $f_n^\beta(x)$ by $u(x, y)$:

$$x\frac{\partial^2 u}{\partial x^2} + \left(1-\beta-y\frac{\partial u}{\partial y}-x\right)\frac{\partial u}{\partial x} + y\frac{\partial u}{\partial y} = 0. \tag{4}$$

We rewrite (4) in the form

$$x\frac{\partial^2 u}{\partial x^2}-y\frac{\partial^2 u}{\partial y\,\partial x}+(1-\beta-x)\frac{\partial u}{\partial x}+y\frac{\partial u}{\partial y}=0,$$

and let

$$L=x\frac{\partial^2}{\partial x^2}-y\frac{\partial^2}{\partial y\,\partial x}+(1-\beta-x)\frac{\partial}{\partial x}+y\frac{\partial}{\partial y}.$$

Also let $A=y\frac{\partial}{\partial y}$. By methods explained in Chapter II we determine operators B and C such that

$$B[f_\nu^\beta(x)\,y^\nu]=f_{\nu-1}^\beta(x)\,y^{\nu-1}$$

and

$$C[f_\nu^\beta(x)\,y^\nu]=-(\nu+1)\,f_{\nu+1}^\beta(x)\,y^{\nu+1}$$

where

$$B=y^{-1}\frac{\partial}{\partial x} \quad \text{and} \quad C=x\,y\frac{\partial}{\partial x}-y^2\frac{\partial}{\partial y}-x\,y-\beta\,y.$$

We find that the operators A, B, and C satisfy the commutator relations $[A, B]=-B$, $[A, C]=C$, $[B, C]=-1$. Therefore, the operators $1, A, B, C$ generate a Lie group. We find also that L satisfies the operator identity

$$L=CB+A.$$

By using this identity and the commutator relations we prove that L commutes with A, B, and C. Therefore, L commutes with

$$R=r_1\,A+r_2\,B+r_3\,C+r_4,$$

where each r_i $(i=1, 2, 3, 4)$ is an arbitrary constant.

We express the extended form of the group generated by the operators B and C as follows:

$$e^{bB}\,f(x, y)=f(x+b\,y^{-1}, y) \tag{5}$$

and

$$e^{cC}\,f(x, y)=e^{-cxy}(1+c\,y)^{-\beta}\,f\big(x(1+c\,y), y(1+c\,y)^{-1}\big), \tag{6}$$

where $f(x, y)$ is an arbitrary function.

A solution of the system

$$Lu=0 \quad \text{and} \quad (A-\nu)\,u=0$$

is the function $f_\nu^\beta(x)\,y^\nu$. Since B and C commute with L, we have

$$L\,e^{cC}\,e^{bB}[f_\nu^\beta(x)\,y^\nu]=e^{cC}\,e^{bB}\,L[f_\nu^\beta(x)\,y^\nu]=0.$$

Therefore, the transformed function, $e^{cC} e^{bB}[f_\nu^\beta(x)\, y^\nu]$, is also annulled by L. We consider three cases:

Case 1. $b=1,\ c=0$,

Case 2. $b=0,\ c=1$.

Case 3. $bc \neq 0$.

Case 1. From (5) we know that for an arbitrary function $f(x, y)$

$$e^B f(x, y) = f(x+y^{-1}, y);$$

then it follows that

$$e^B [f_\nu^\beta(x)\, y^\nu] = y^\nu f_\nu^\beta(x+y^{-1}).$$

We have thus obtained the generating function which has the following expansion:

$$y^\nu f_\nu^\beta(x+y^{-1}) = \sum_{n=0}^{\nu} \frac{1}{n!} f_{\nu-n}^\beta(x)\, y^{\nu-n}.$$

If we divide by y^ν and let $t=y^{-1}$, we get

$$f_\nu^\beta(x+t) = \sum_{n=0}^{\nu} \frac{1}{n!} f_{\nu-n}^\beta(x)\, t^n. \tag{7}$$

Case 2. From (6) we know that

$$e^C f(x, y) = e^{-xy}(1+y)^{-\beta} f\big(x(1+y), y(1+y)^{-1}\big),$$

for an arbitrary function $f(x, y)$; then

$$e^C [f_\nu^\beta(x)\, y^\nu] = e^{-xy}(1+y)^{-\beta} \left(\frac{y}{1+y}\right)^\nu f_\nu^\beta\big(x(1+y)\big).$$

From the expansion of this function we obtain the following generating relation, where $t=-y$:

$$e^{xt}(1-t)^{-\beta-\nu} f_\nu^\beta\big(x(1-t)\big) = \sum_{n=0}^{\infty} \frac{(\nu+1)_n}{n!} f_{\nu+n}^\beta(x)\, t^n. \tag{8}$$

If $\nu=0$, this relation becomes

$$e^{xt}(1-t)^{-\beta} = \sum_{n=0}^{\infty} f_n^\beta(x)\, t^n. \tag{9}$$

Case 3. For $bc \neq 0$ we choose $c=1$ and $b=-\dfrac{1}{w}$. Then for an arbitrary function $f(x, y)$ we have

$$\begin{aligned} e^C e^{-\frac{1}{w}B} f(x, y) &= e^C f\left(x - \frac{1}{wy}, y\right) \\ &= e^{-xy}(1+y)^{-\beta} f\left(\left(x - \frac{1}{wy}\right)(1+y), y(1+y)^{-1}\right). \end{aligned}$$

Hence

$$e^C e^{-\frac{1}{w}B}[f_\nu^\beta(x)\, y^\nu] = e^{-xy}(1+y)^{-\beta}[y(1+y)^{-1}]^\nu f_\nu^\beta\left(\left(x-\frac{1}{wy}\right)(1+y)\right).$$

If in this generating function we let $y=-t$ and use the identity

$$y^{-\nu} f_\nu^\beta(y) = \frac{1}{\Gamma(\nu+1)}\, {}_2F_0\left[\begin{matrix}-\nu, & \beta; & \dfrac{-1}{y}\\ & -; & \end{matrix}\right] \tag{10}$$

for simplification purposes, we obtain the following expansion:

$$\begin{aligned} &e^{xt}(1-t)^{-\beta}(1+xwt)^\nu\, {}_2F_0\left[\begin{matrix}-\nu, & \beta; & \dfrac{-wt}{(1+xwt)(1-t)}\\ & -; & \end{matrix}\right] \\ &\quad = \sum_{n=0}^{\infty} {}_2F_0\left[\begin{matrix}-n, & -\nu; & w\\ & -; & \end{matrix}\right] f_n^\beta(x)\, t^n. \end{aligned} \tag{11}$$

If in (11) we replace $-\nu$ by c, then for arbitrary c we have the (divergent) generating relation:

$$\begin{aligned} &e^{xt}(1-t)^{-\beta}(1+xwt)^{-c}\, {}_2F_0\left[\begin{matrix}c, & \beta; & \dfrac{-wt}{(1+xwt)(1-t)}\\ & -; & \end{matrix}\right] \\ &\quad \cong \sum_{n=0}^{\infty} {}_2F_0\left[\begin{matrix}-n, & c; & w\\ & -; & \end{matrix}\right] f_n^\beta(x)\, t^n. \end{aligned} \tag{12}$$

The generating function of (12) is obtained in Chapter I, Section 6, by using series manipulation and in Chapter V, Section 6, by using Brafman's contour integral method with Chaundy's formula.

If we let $S=e^{cC}e^{bB}$, where b and c are arbitrary constants, we find that

$$\begin{aligned} S(A-\nu)S^{-1} &= e^{cC}e^{bB}(A-\nu)e^{-bB}e^{-cC} \\ &= e^{cC}[A+bB-\nu]\,e^{-cC} \\ &= A-cC+b(B+c)-\nu \\ &= A+bB-cC+bc-\nu. \end{aligned}$$

Let $R=r_1A+r_2B+r_3C+r_4$, where r_1, r_2, r_3, r_4 are constants and r_1, r_2, r_3 are not all zero. For all possible choices of b and c we see that $A-\nu$ is not conjugate to any R for which $r_1=0$. Hence solutions of the system

$$Lu=0 \quad \text{and} \quad (r_2B+r_3C+r_4)\,u=0,$$

where r_2 and r_3 are not both zero, have not been included in the consideration of this section.

3. The simple Bessel polynomials. Krall and Frink [1] studied the simple Bessel polynomials defined as follows:

$$f_n(x) = {}_2F_0\left(-n, n+1; -; \frac{-x}{2}\right), \qquad n > 0,$$

with $f_{-n}(x) = f_{n-1}(x)$ and $f_{-1}(x) = f_0(x) = 1$. These polynomials satisfy the two differential recurrence relations

$$x^2 f_n'(x) = f_{n-1}(x) + (nx - 1) f_n(x), \tag{1}$$

$$x^2 f_n'(x) = f_{n+1}(x) - [1 + (n+1)x] f_n(x), \tag{2}$$

and hence

$$x^2 f_n''(x) + (2x+2) f_n'(x) - n(n+1) f_n(x) = 0. \tag{3}$$

From (3) we construct a partial differential equation, replacing $\frac{d}{dx}$ by $\frac{\partial}{\partial x}$, n by $y\frac{\partial}{\partial y}$, and $f_n(x)$ by $u(x, y)$:

$$x^2 \frac{\partial^2 u}{\partial x^2} + (2x+2)\frac{\partial u}{\partial x} - y\frac{\partial}{\partial y}\left(y\frac{\partial}{\partial y} + 1\right) u = 0. \tag{4}$$

Let

$$L = x^2 \frac{\partial^2}{\partial x^2} + (2x+2)\frac{\partial}{\partial x} - y\frac{\partial}{\partial y}\left(y\frac{\partial}{\partial y} + 1\right)$$

$$= x^2 \frac{\partial^2}{\partial x^2} - y^2 \frac{\partial^2}{\partial y^2} + (2x+2)\frac{\partial}{\partial x} - 2y\frac{\partial}{\partial y}.$$

Using the technique described in detail in Chapter II, we determine with the aid of (1) and (2) the differential operators B and C such that

$$B[f_\nu(x)\, y^\nu] = f_{\nu-1}(x)\, y^{\nu-1} \tag{5}$$

and

$$C[f_\nu(x)\, y^\nu] = f_{\nu+1}(x)\, y^{\nu+1}, \tag{6}$$

where

$$B = x^2 y^{-1}\frac{\partial}{\partial x} - x\frac{\partial}{\partial y} + y^{-1}$$

and

$$C = x^2 y\frac{\partial}{\partial x} + xy^2\frac{\partial}{\partial y} + xy + y.$$

Let $A = y\frac{\partial}{\partial y}$. The operators A, B, and C satisfy the commutator relations $[A, B] = -B$, $[A, C] = C$, $[B, C] = 0$. Therefore, these operators generate a three-parameter Lie group.

From the computation of $[B, C]$, we find that

$$x^2 L = BC - 1.$$

By using the commutator relations we prove that the operators A, B, C commute with $x^2 L$.

We express the extended form of the group generated by each of the operators A, B, C as follows:

$$e^{aA} f(x, y) = f(x, e^a y), \tag{7}$$

$$e^{bB} f(x, y) = \exp\left\{\frac{1-\sqrt{1-2bxy^{-1}}}{x}\right\} \cdot f\left(\frac{x}{\sqrt{1-2bxy^{-1}}}, y\sqrt{1-2bxy^{-1}}\right), \tag{8}$$

$$e^{cC} f(x, y) = (1-2cxy)^{-\frac{1}{2}} \exp\left\{\frac{1-\sqrt{1-2cxy}}{x}\right\} \cdot f\left(\frac{x}{\sqrt{1-2cxy}}, \frac{y}{\sqrt{1-2cxy}}\right), \tag{9}$$

where a, b, c are arbitrary constants and $f(x, y)$ is an arbitrary function. Since $[B, C]=0$, we know that B and C commute and that we may write

$$e^{bB} e^{cC} f(x, y) = e^{bB+cC} f(x, y) = (1-2cy)^{-\frac{1}{2}} \exp\left\{\frac{1-\sqrt{1-2bxy^{-1}}\sqrt{1-2cxy}}{x}\right\} f(\xi, \eta) \tag{10}$$

where $\xi = \dfrac{x}{\sqrt{1-2bxy^{-1}}\sqrt{1-2cxy}}$ and $\eta = \dfrac{y\sqrt{1-2bxy^{-1}}}{\sqrt{1-2cxy}}$. From (7) we see that A generates a trivial group.

Since $y^\nu f_\nu(x)$ is a solution of the system

$$Lu = 0 \quad \text{and} \quad (A-\nu)u = 0,$$

we determine generating functions of $\{f_n(x)\}$ by finding $e^{bB+cC}[y^\nu f_\nu(x)]$. We need to consider three cases:

Case 1. $b=1, c=0$,

Case 2. $b=0, c=1$,

Case 3. $bc \neq 0$.

Case 1. Since for arbitrary $f(x, y)$

$$e^{B} f(x, y) = \exp\left\{\frac{1-\sqrt{1-2xy^{-1}}}{x}\right\} f\left(\frac{x}{\sqrt{1-2xy^{-1}}}, \sqrt{y(y-2x)}\right),$$

we find

$$e^{B}[y^\nu f_\nu(x)] = \exp\left\{\frac{1-\sqrt{1-2xy^{-1}}}{x}\right\} [y(y-2x)]^{\nu/2} f_\nu\left(\frac{x}{\sqrt{1-2xy^{-1}}}\right).$$

Also since $B[y^\nu f_\nu(x)] = y^{\nu-1} f_{\nu-1}(x)$, we have

$$e^B[y^\nu f_\nu(x)] = \sum_{n=0}^{\infty} \frac{1}{n!} f_{\nu-n}(x)\, y^{\nu-n}.$$

By equating the two expressions for $e^B[y^\nu f_\nu(x)]$, dividing by y^ν and replacing y^{-1} by t, we get the generating relation

$$(1-2xt)^{\nu/2} \exp\left\{\frac{1-\sqrt{1-2xt}}{x}\right\} f_\nu\left(\frac{x}{\sqrt{1-2xt}}\right) = \sum_{n=0}^{\infty} \frac{1}{n!} f_{\nu-n}(x)\, t^n,$$

or equivalently,

$$(1-2xt)^{\nu/2} \exp\left\{\frac{2t}{1+\sqrt{1-2xt}}\right\} f_\nu\left(\frac{x}{\sqrt{1-2xt}}\right) = \sum_{n=0}^{\infty} \frac{1}{n!} f_{\nu-n}(x)\, t^n. \quad (11)$$

Case 2. From (6) and (9) we obtain the generating relation

$$(1-2xy)^{-\frac{(\nu+1)}{2}} \exp\left\{\frac{1-\sqrt{1-2xy}}{x}\right\} f_\nu\left(\frac{x}{\sqrt{1-2xy}}\right) = \sum_{n=0}^{\infty} \frac{1}{n!} f_{\nu+n}(x)\, y^n. \quad (12)$$

Since $f_{-1}(x) = 1$, we get for $\nu = -1$

$$\exp\left\{\frac{1-\sqrt{1-2xy}}{x}\right\} = \sum_{n=0}^{\infty} \frac{1}{n!} f_{n-1}(x)\, y^n. \quad (12a)$$

See Krall and Frink [1; p. 106].

Also since $f_0(x) = 1$, we get for $\nu = 0$

$$(1-2xy)^{-\frac{1}{2}} \exp\left\{\frac{1-\sqrt{1-2xy}}{x}\right\} = \sum_{n=0}^{\infty} \frac{1}{n!} f_n(x)\, y^n, \quad (12b)$$

or equivalently,

$$(1-2xy)^{-\frac{1}{2}} \exp\left\{\frac{2y}{1+\sqrt{1-2xy}}\right\} = \sum_{n=0}^{\infty} \frac{1}{n!} f_n(x)\, y^n.$$

See Burchnall [1].

Using $f_{-n}(x) = f_{n-1}(x)$, we can establish the equivalence of the generating relations obtained in Case 1 and Case 2.

Case 3. From (10) with $b = -\frac{1}{2}w$ and $c = \frac{1}{2}w$, where $w \neq 0$, we get the following bilateral generating relation:

$$\begin{aligned}(1+wxy^{-1})^{\frac{\nu}{2}} (1-wxy)^{-\frac{\nu+1}{2}} \\ \cdot \exp\left\{\frac{w(y-y^{-1}+wx)}{1+\sqrt{1-wxy}\sqrt{1+wxy^{-1}}}\right\} f_\nu\left(\frac{x}{\sqrt{1-wxy}\sqrt{1+wxy^{-1}}}\right) \\ = \sum_{n=-\infty}^{\infty} J_n(w) f_{n+\nu}(x)\, y^n.\end{aligned} \quad (13)$$

Let $S=e^{bB+cC}$. Then $S(A-\nu)S^{-1}=A+bB-cC-\nu$. Let

$$R=r_1 A+r_2 B+r_3 C+r_4.$$

We see that $A-\nu$ is not conjugate to R if $r_1=0$. Hence solutions of the system $Lu=0$ and $(r_2 B+r_3 C+r_4)u=0$, where r_2 and r_3 are not both zero, have not been included in the consideration of this section.

We now list the generating functions which we have obtained for the simple Bessel polynomials:

$$(1-2xy)^{-\frac{\nu+1}{2}}\exp\left\{\frac{1-\sqrt{1-2xy}}{x}\right\}f_\nu\left(\frac{x}{\sqrt{1-2xy}}\right)=\sum_{n=0}^{\infty}\frac{1}{n!}f_{\nu+n}(x)\,y^n, \tag{12}$$

$$\exp\left\{\frac{1-\sqrt{1-2xy}}{x}\right\}=\sum_{n=0}^{\infty}\frac{1}{n!}f_{n-1}(x)\,y^n, \tag{12a}$$

$$(1-2xy)^{-\frac{1}{2}}\exp\left\{\frac{1-\sqrt{1-2xy}}{x}\right\}=\sum_{n=0}^{\infty}\frac{1}{n!}f_n(x)\,y^n, \tag{12b}$$

$$\begin{aligned}&(1+wxy^{-1})^{\frac{\nu}{2}}(1-wxy)^{-\frac{\nu+1}{2}}\\ &\quad\cdot\exp\left\{\frac{w(y-y^{-1}+wx)}{1+\sqrt{1-wxy}\sqrt{1+wxy}}\right\}f_\nu\left(\frac{x}{\sqrt{1-wxy}\sqrt{1+wxy^{-1}}}\right)\\ &\quad=\sum_{n=-\infty}^{\infty}J_n(w)\,f_{n+\nu}(x)\,y^n.\end{aligned} \tag{13}$$

4. The Gegenbauer polynomials. The Gegenbauer polynomials of degree n may be defined as follows:

$$C_n^\nu(x)=\sum_{k=0}^{[n/2]}\frac{(-1)^k(\nu)_{n-k}(2x)^{n-2k}}{k!(n-2k)!}.$$

In order to find generating functions for the set $\{C_n^\nu(x)\}$ we begin with two independent recurrence relations satisfied by each element of this set:

$$(1-x^2)DC_n^\nu(x)=(2\nu-1+n)\,C_{n-1}^\nu(x)-nx\,C_n^\nu(x) \tag{1}$$

and

$$(1-x^2)DC_n^\nu(x)=-(n+1)\,C_{n+1}^\nu(x)+(2\nu+n)\,x\,C_n^\nu(x), \tag{2}$$

where $D=d/dx$. From (1) and (2) we determine the differential equation

$$(1-x^2)D^2C_n^\nu(x)-(2\nu+1)\,xDC_n^\nu(x)+n(2\nu+n)\,C_n^\nu(x)=0. \tag{3}$$

From (3) we construct a partial differential equation, replacing D by $\frac{\partial}{\partial x}$, n by $y\frac{\partial}{\partial y}$, and $C_n^\nu(x)$ by $u(x,y)$:

$$(1-x^2)\frac{\partial^2 u}{\partial x^2}-(2\nu+1)\,x\frac{\partial u}{\partial x}+y\frac{\partial}{\partial x}+y\frac{\partial}{\partial y}\left(2\nu+y\frac{\partial}{\partial y}\right)u=0. \tag{4}$$

We may rewrite (4) in the following form:

$$(1-x^2)\frac{\partial^2 u}{\partial x^2}+y^2\frac{\partial^2 u}{\partial y^2}-(2\nu+1)\,x\frac{\partial u}{\partial x}+(2\nu+1)\,y\frac{\partial u}{\partial y}=0.$$

Let $A=y\dfrac{\partial}{\partial y}$ and let L represent the differential operator of (4), i.e.,

$$L=(1-x^2)\frac{\partial^2}{\partial x^2}+y^2\frac{\partial^2}{\partial y^2}-(2\nu+1)\,x\frac{\partial}{\partial x}+(2\nu+1)\,y\frac{\partial}{\partial y}.$$

We determine operators B and C such that

$$B[C_n^\nu(x)\,y^n]=-(2\nu-1+n)\,C_{n-1}^\nu(x)\,y^{n-1}$$

and

$$C[C_n^\nu(x)\,y^n]=(n+1)\,C_{n+1}^\nu(x)\,y^{n+1},$$

where

$$B=(x^2-1)\,y^{-1}\frac{\partial}{\partial x}-x\frac{\partial}{\partial y}$$

and

$$C=(x^2-1)\,y\frac{\partial}{\partial x}+x\,y^2\frac{\partial}{\partial y}+2\nu\,x\,y.$$

The operators A, B, C satisfy the following commutator relations: $[A,B]=-B$, $[A,C]=C$, $[B,C]=-2A-2\nu$. Therefore, the operators 1, A, B, C generate a Lie group.

The second order differential operator L satisfies the operator identity

$$(1-x^2)L=CB+A^2+(2\nu-1)A.$$

By means of this identity and the commutator relations we prove that $(1-x^2)L$ commutes with each of the operators A, B, C. Then for arbitrary constants b and c the operator $e^{cC}e^{bB}$ will transform solutions of L into solutions of L; in other words,

$$e^{cC}e^{bB}(1-x^2)\,Lu=(1-x^2)\,L(e^{cC}e^{bB}u)=0$$

if and only if $Lu=0$.

The extended forms of the transformation groups generated by the operators B and C may be expressed as follows:

$$e^{bB}f(x,y)=f\left(\frac{x\,y-b}{\sqrt{y^2-2b\,x\,y+b^2}},\sqrt{y^2-2b\,x\,y+b^2}\right)$$

and

$$e^{cC}f(x,y)=(c^2y^2-2c\,x\,y+1)^{-\nu}$$
$$\cdot f\left(\frac{x-c\,y}{\sqrt{c^2y^2-2c\,x\,y+1}},\frac{y}{\sqrt{c^2y^2-2c\,x\,y+1}}\right)$$

where b and c are arbitrary constants and $f(x, y)$ is an arbitrary function. We find

$$e^{cC} e^{bB}[C_k^\nu(x)\, y^k] = (c^2 y^2 - 2cxy + 1)^{-\nu} C_k^\nu(\xi)\, \eta^k, \tag{5}$$

where

$$\xi = \frac{-cy^2(1+bc) + xy(1+2bc) - b}{\sqrt{c^2y^2 - 2cxy + 1}\sqrt{y^2(1+bc)^2 - 2bxy(1+bc) + b^2}}$$

and

$$\eta = \frac{\sqrt{(1+bc)^2 y^2 - 2bxy(1+bc) + b^2}}{\sqrt{1 - 2cxy + c^2y^2}}.$$

Part I. *Generating functions derived from the operator* $(A-\nu)$. We will consider three cases of the transformed function $e^{cC} e^{bB}[C_k^\nu(x)\, y^k]$:

Case 1. $b=1$, $c=0$,

Case 2. $b=0$, $c=1$,

Case 3. $bc \neq 0$.

Part I, *Case* 1. If we choose $b=1$ and $c=0$, we find

$$e^B[y^k C_k^\nu(x)] = (y^2 - 2xy + 1)^{k/2} C_k^\nu\left(\frac{xy-1}{\sqrt{y^2-2xy+1}}\right).$$

By expanding this function we get

$$(y^2 - 2xy + 1)^{k/2} C_k^\nu\left(\frac{xy-1}{\sqrt{y^2-2xy+1}}\right) = \sum_{n=0}^{k} \frac{(1-2\nu-k)_n}{n!} C_{k-n}^\nu(x)\, y^{k-n}.$$

If we divide by y^k and let $t = y^{-1}$, we get

$$(1 - 2xt + t^2)^{k/2} C_k^\nu\left(\frac{x-t}{\sqrt{1-2xt+t^2}}\right) = \sum_{n=0}^{k} \frac{(1-2\nu-k)_n}{n!} C_{k-n}^\nu(x)\, t^n. \tag{6}$$

Part I, *Case* 2. If we choose $b=0$ and $c=1$, we get

$$e^C[y^k C_k^\nu(x)] = y^k (y^2 - 2xy + 1)^{-\nu-k/2} C_k^\nu\left(\frac{x-y}{\sqrt{y^2-2xy+1}}\right).$$

If we expand this function and divide by y^k, we get the generating relation

$$(y^2 - 2xy + 1)^{-\nu-k/2} C_k^\nu\left(\frac{x-y}{\sqrt{y^2-2xy+1}}\right) = \sum_{n=0}^{\infty} \binom{k+n}{n} C_{k+n}^\nu(x)\, y^n. \tag{7}$$

See Rainville [1; p. 280, (23)].

Part I, *Case* 3. For $bc \neq 0$ we choose $b=-1$ and $c=1$. This choice is suggested by the frequency of occurrence in (5) of the factor $1+bc$.

We then have

$$e^{C} e^{-B}[C_k^{\nu}(x)\, y^k] = (y^2 - 2xy + 1)^{-\nu}\, C_k^{\nu}(\xi)\, \eta^k,$$

where

$$\xi = \frac{1-xy}{\sqrt{1-2xy+y^2}} \quad \text{and} \quad \eta = \frac{1}{\sqrt{1-2xy+y^2}}.$$

We expand this generating function as follows:

$$(1-2xy+y^2)^{-\nu-k/2}\, C_k^{\nu}\left(\frac{1-xy}{\sqrt{1-2xy+y^2}}\right) = \sum_{n=0}^{\infty} \frac{(2\nu+n)_k}{k!}\, C_n^{\nu}(x)\, y^n. \quad (8)$$

If we let $\rho = \sqrt{1-2xy+y^2}$ we may rewrite (8) in the form

$$\rho^{-2\nu-k}\, C_k^{\nu}\left(\frac{1-xy}{\rho}\right) = \sum_{n=0}^{\infty} \frac{(2\nu+n)_k}{k!}\, C_n^{\nu}(x)\, y^n.$$

In order to express the left member of (8) in hypergeometric form we use

$$C_n^{\nu}(x) = \frac{(2\nu)_n\, x^n}{n!}\; {}_2F_1\left[\begin{matrix} -\tfrac{1}{2}n, & -\tfrac{1}{2}n+\tfrac{1}{2}; & \dfrac{x^2-1}{x^2} \\ & \nu+\tfrac{1}{2}; & \end{matrix}\right].$$

See Rainville [1; p. 280, (20)]. Then after some simplification, Eq. (8) yields

$$\begin{aligned} &(1-2xy+y^2)^{-\nu-k}(1-xy)^k\; {}_2F_1\left[\begin{matrix} -\tfrac{1}{2}k, & -\tfrac{1}{2}k+\tfrac{1}{2}; & \dfrac{y^2(x^2-1)}{(1-xy)^2} \\ & \nu+\tfrac{1}{2}; & \end{matrix}\right] \\ &\quad = \sum_{n=0}^{\infty} \frac{(2\nu+k)_n}{(2\nu)_n}\, C_n^{\nu}(x)\, y^n. \end{aligned} \quad (9)$$

We will now apply to the hypergeometric function in the left member of (9) the following theorem:

"If $|z|<1$, $F(a,b;c;z) = (1-z)^{c-a-b}\, F(c-a, c-b; c; z)$."

See Rainville [1, p. 60, Theorem 21]. By applying this theorem and letting $\gamma = 2\nu + k$, from (9) we obtain

$$(1-xy)^{-\gamma}\; {}_2F_1\left[\begin{matrix} \tfrac{1}{2}\gamma, & \tfrac{1}{2}\gamma+\tfrac{1}{2}; & \dfrac{y^2(x^2-1)}{(1-xy)^2} \\ & \nu+\tfrac{1}{2}; & \end{matrix}\right] = \sum_{n=0}^{\infty} \frac{(\gamma)_n}{(2\nu)_n}\, C_n^{\nu}(x)\, y^n. \quad (10)$$

The generating relation (10) is found by Rainville [1; p. 279, (8)] by means of series manipulations.

Part II. *Generating functions derived from operators not conjugate to* $(A-\nu)$. The three generating functions of (6), (7), and (8) have been obtained by transforming $C_k^{\nu}(x)\, y^k$, which is a solution of

$$Lu = 0 \quad \text{and} \quad (A-k)\,u = 0.$$

If we wish to obtain additional generating functions for the Gegenbauer polynomials, we need to find operators which are not conjugate to $A-k$; i.e., we wish to find first order differential operators R such that for all choices of b and c

$$S(A-k)S^{-1} \neq R, \quad \text{where} \quad S = e^{cC} e^{bB}.$$

Consider the set of linear differential operators

$$\{R | R = r_1 A + r_2 B + r_3 C + r_4,$$

for all combinations of zero and nonzero coefficients except for

$$r_1 = r_2 = r_3 = 0\}.$$

We find that

$$S(A-k)S^{-1} = (1+2bc)A + bB - c(1+bc)C + (2vbc-k).$$

Then for $r_1 = 1+2bc$, $r_2 = b$, $r_3 = -c(1+bc)$, we have $r_1^2 + 4r_2 r_3 = 1$. Therefore, $A-k$ is not conjugate to operators for which $r_1^2 + 4r_2 r_3 = 0$. We consider each of the following cases:

Case 1. $r_1 = 0$, $r_2 \neq 0$, $r_3 = 0$,

Case 2. $r_1 = 2$, $r_2 = 1$, $r_3 = -1$,

Case 3. $r_1 = 0$, $r_2 = 0$, $r_3 \neq 0$.

Part II, *Case* 1. If $r_1 = 0$, $r_2 = 1$, $r_3 = 0$, we seek a solution of the system

$$Lu = 0 \quad \text{and} \quad (B+1)u = 0.$$

A solution of this system is

$$u(x,y) = e^{xy} {}_0F_1\left[\begin{matrix} -; \\ v+\frac{1}{2}; \end{matrix} \quad \frac{y^2(x^2-1)}{4}\right].$$

If we expand this function, we get

$$e^{xy} {}_0F_1\left[\begin{matrix} -; \\ v+\frac{1}{2}; \end{matrix} \quad \frac{y^2(x^2-1)}{4}\right] = \sum_{n=0}^{\infty} \frac{C_n^v(x)}{(2v)_n} y^n. \tag{11}$$

See Rainville [1; p. 278, (7)].

Part II, *Case* 2. For $r_1 r_2 r_3 \neq 0$ and $r_1^2 + 4r_2 r_3 = 0$ we choose $r_1 = 2$, $r_2 = 1$, $r_3 = -1$. We are led to this choice by considering $e^{cC}(B+w)e^{-cC}$, where w is a nonzero constant. We find that

$$e^{cC}(B+w)e^{-cC} = 2cA + B - c^2 C + (2cv + w).$$

If we let $c = 1$, we get $r_1 = 2$, $r_2 = 1$, $r_3 = -1$. Since this choice satisfies the above conditions, we may determine a solution of the system

$$Lu = 0 \quad \text{and} \quad (2A + B - C + 2v + w)u = 0$$

from the generating function of (11) by replacing y by $w\,y$ and transforming as follows:

$$e^C\,e^{wxy}\,{}_0F_1\left[\begin{matrix}-; & \dfrac{w^2\,y^2(x^2-1)}{4}\\ \nu+\frac{1}{2}; & \end{matrix}\right]$$

$$=\rho^{-2\nu}\exp\left\{\frac{w\,y(x-y)}{\rho^2}\right\}{}_0F_1\left[\begin{matrix}-; & \dfrac{w^2\,y^2(x^2-1)}{4\rho^4}\\ \nu+\frac{1}{2}; & \end{matrix}\right],$$

where $\rho=(1-2x\,y+y^2)^{\frac{1}{2}}$. When this function is expanded we obtain the following bilateral generating relation:

$$\rho^{-2\nu}\exp\left\{\frac{w\,y(x-y)}{\rho^2}\right\}{}_0F_1\left[\begin{matrix}-; & \dfrac{w^2\,y^2(x^2-1)}{4\rho^4}\\ \nu+\frac{1}{2}; & \end{matrix}\right]$$
$$=\sum_{n=0}^{\infty}\frac{n!}{(2\nu)_n}\,L_n^{2\nu-1}(-w)\,C_n^\nu(x)\,y^n. \tag{12}$$

See Rainville [1; p. 281, (24)].

Part II, *Case* 3. For $r_1=0$, $r_2=0$, and $r_3\neq 0$, we seek a solution of the system

$$L u=0 \qquad (C+\lambda)\,u=0,$$

where λ is an arbitrary constant. We may avoid actually solving this system by noting that

$$e^{bB}\,e^{cC}(B+1)\,e^{-cC}\,e^{-bB}$$
$$=2c(1+b\,c)\,A+(1+b\,c)^2-c^2\,C+2\nu\,c(1+b\,c)+1.$$

If we choose $b=1$ and $c=-1$, we get

$$e^B\,e^{-C}(B+1)\,e^C\,e^{-B}=-C+1.$$

Therefore, we can obtain a solution of $L u=0$ and $(C-1)\,u=0$, by transforming the generating function of (11) as follows:

$$e^B\,e^{-C}\,e^{xy}\,{}_0F_1\left[\begin{matrix}-; & \dfrac{y^2(x^2-1)}{4}\\ \nu+\frac{1}{2}; & \end{matrix}\right]$$

$$=y^{-2\nu}\exp\left\{\frac{y-x}{y}\right\}{}_0F_1\left[\begin{matrix}-; & \dfrac{x^2-1}{4y^2}\\ \nu+\frac{1}{2}; & \end{matrix}\right].$$

If we let $t=-1/y$, we get

$$e(-t)^{2\nu}\,e^{xt}\,{}_0F_1\left[\begin{matrix}-; & \dfrac{t^2(x^2-1)}{4}\\ \nu+\frac{1}{2}; & \end{matrix}\right]$$

as our generating function. But this function differs only trivially from (11).

We have now obtained for the Gegenbauer polynomials the following generating relations:

$$\rho^k\, C_k^\nu\left(\frac{x-y}{\rho}\right)=\sum_{n=0}^{k}\frac{(1-2\nu-k)_n}{n!}\, C_{k-n}^\nu\, y^n, \tag{6}$$

$$\rho^{-2\nu-k}\, C_k^\nu\left(\frac{x-y}{\rho}\right)=\sum_{n=0}^{\infty}\binom{k+n}{n}\, C_{k+n}^\nu(x)\, y^n, \tag{7}$$

$$\rho^{-2\nu-k}\, C_k^\nu\left(\frac{1-xy}{\rho}\right)=\sum_{n=0}^{\infty}\frac{(2\nu+n)_k}{k!}\, C_n^\nu(x)\, y^n, \tag{8}$$

$$e^{xy}\,{}_0F_1\left[\begin{matrix}-; \\ \nu+\frac{1}{2};\end{matrix}\ \frac{y^2(x^2-1)}{4}\right]=\sum_{n=0}^{\infty}\frac{C_n^\nu(x)}{(2\nu)_n}\, y^n, \tag{11}$$

$$\begin{aligned}\rho^{-2\nu}\exp\left\{\frac{-wy(x-y)}{\rho^2}\right\}{}_0F_1\left[\begin{matrix}-; \\ \nu+\frac{1}{2};\end{matrix}\ \frac{w^2y^2(x^2-1)}{4\rho^4}\right]\\ =\sum_{n=0}^{\infty}\frac{n!}{(2\nu)_n}\, L_n^{2\nu-1}(w)\, C_n^\nu(x)\, y^n,\end{aligned} \tag{12}$$

where $\rho=\sqrt{1-2xy+y^2}$.

B. Viswanathan [1], whose doctoral research was directed by Professor Weisner, used both first and second order differential operators to find generating functions for the ultraspherical function $P_n^\lambda(x)$. (If n is a positive integer, this function is a polynomial identical with the Gegenbauer polynomial $C_n^\lambda(x)$.)

Willard Miller, Jr., [1] gives further insight into the Weisner method in his work which relates Lie groups and special functions.

Chapter IV

The Truesdell Method

1. Introduction. Truesdell's study of the functional equation

$$\frac{\partial}{\partial z} F(z, \alpha) = F(z, \alpha + 1),$$

called the F-equation, yielded many valuable results, among which is his method for obtaining generating functions. By this method Truesdell [1] obtained both ascending and descending generating functions. For example, if the elements of a given set of functions $\{\phi_n(x)\}$ satisfy a differential-difference equation of the ascending type,

$$\phi'_n(x) = A(x, n)\, \phi_n(x) + B(x, n)\, \phi_{n+1}(x),$$

with coefficients $A(x, n)$ and $B(x, n)$ suitably restricted, then a generating function can be found for the set $\{\phi_{k+n}(x)\}$, where k is fixed. Furthermore, if the elements of the set satisfy a differential-difference equation of the descending type,

$$\phi'_n(x) = C(x, n)\, \phi_n(x) + D(x, n)\, \phi_{n-1}(x),$$

with coefficients suitably restricted, a generating function can be found for the set $\{\phi_{k-n}(x)\}$, when k is fixed.

The reader will note that in using either the Truesdell method or the Weisner method we begin with two differential-difference equations, one of ascending type and one of descending type. In the Truesdell method we make independent use of each of these two equations, but in the Weisner method we use them simultaneously.

2. The ascending equation. We begin our study of the Truesdell method by considering transformations leading to the F-equation,

$$\frac{\partial}{\partial z} F(z, \alpha) = F(z, \alpha + 1).$$

Before giving a detailed explanation of the method we will indicate the basic steps in Eqs. (1), (2), (3), and (4) below.

Suppose we wish to use this method to obtain a generating function for a set of functions $\{\phi_n(x)\}$. Using Truesdell's notation we let $f(y,\alpha)=\phi_\alpha(y)$. Then if $f(y,\alpha)$ satisfies the ascending equation

$$\frac{\partial}{\partial y} f(y,\alpha)=A(y,\alpha)\, f(y,\alpha)+B(y,\alpha)\, f(y,\alpha+1), \tag{1}$$

where $B(y,\alpha)\neq 0$, it is always possible to transform $f(y,\alpha)$ into a function $g(y,\alpha)$ which satisfies the equation

$$\frac{\partial}{\partial y} g(y,\alpha)=C(y,\alpha)\, g(y,\alpha+1). \tag{2}$$

Furthermore, if $C(y,\alpha)$ satisfies the factorability condition

$$C(y,\alpha)=A(\alpha)\, Y(y),$$

then $g(y,\alpha)$ is transformable into a function $F(z,\alpha)$ such that

$$\frac{\partial F(z,\alpha)}{\partial z}=F(z,\alpha+1). \tag{3}$$

Finally, if $F(z+y,\alpha)$ possesses a Maclaurin series in powers of y, a generating relation may be expressed formally as follows:

$$F(z+y,\alpha)=\sum_{n=0}^{\infty} \frac{y^n}{n!} F(z,\alpha+n). \tag{4}$$

For convenience we will call (1) an f-type equation, (2) a g-type equation, and (3) the F-equation.

We will now explain how, in general, the two transformations, from (1) to (2) and from (2) to (3), can be accomplished. Assume that for a given set of functions $\{\phi_n(x)\}$ we are able to determine at least one differential difference equation of the following type:

$$\phi_n'(x)=A(x,n)\,\phi_n(x)+B(x,n)\,\phi_{n+1}(x), \tag{5}$$

where $B(x,n)\neq 0$. If we let $f(y,\alpha)=\phi_\alpha(y)$, then (5) yields

$$\frac{\partial}{\partial y} f(y,\alpha)=A(y,\alpha)\, f(y,\alpha)+B(y,\alpha)\, f(y,\alpha+1). \tag{6}$$

We now seek a transformation of $f(y,\alpha)$ which will eliminate the term $A(y,\alpha)\, f(y,\alpha)$ in (6). Accordingly, let

$$g(y,\alpha)=\exp\left\{-\int_{y_0}^{y} A(v,\alpha)\, dv\right\} f(y,\alpha), \tag{7}$$

where y_0 is a constant chosen to simplify the form of $g(y, \alpha)$. Then

$$\frac{\partial g(y, \alpha)}{\partial y} = \exp\left\{-\int_{y_0}^{y} A(v, \alpha)\, dv\right\} B(y, \alpha) f(y, \alpha+1). \tag{8}$$

We define the operator symbol $\underset{\alpha}{\Delta}$ as follows:

$$\underset{\alpha}{\Delta} A(v, \alpha) = A(v, \alpha+1) - A(v, \alpha).$$

In order to write the right member of (8) in the form $C(y, \alpha)\, g(y, \alpha+1)$, we use the identity:

$$-\int_{y_0}^{y} A(v, \alpha)\, dv = -\int_{y_0}^{y} A(v, \alpha+1)\, dv + \underset{\alpha}{\Delta} \int_{y_0}^{y} A(v, \alpha)\, dv.$$

Then (8) may be rewritten as

$$\frac{\partial}{\partial y} g(y, \alpha) = B(y, \alpha) \exp\left\{\underset{\alpha}{\Delta} \int_{y_0}^{y} A(v, \alpha)\, dv\right\} g(y, \alpha+1). \tag{9}$$

Let

$$C(y, \alpha) = B(y, \alpha) \exp\left\{\underset{\alpha}{\Delta} \int_{y_0}^{y} A(v, \alpha)\, dv\right\}.$$

If $C(y, \alpha)$ does not satisfy the factorability condition

$$C(y, \alpha) = Y(y)\, A(\alpha),$$

the Truesdell method does not apply. This condition is both necessary and sufficient for an equation of the g-type to be reducible to the F-equation. Truesdell [1; p. 31, Th. 7.1] proves that the factorability condition is equivalent to the following restriction on the coefficients in the f-type ascending equation:

There exists a function $L(y)$ such that

$$\underset{\alpha}{\Delta} A(y, \alpha) + \frac{\partial}{\partial y} \log B(y, \alpha) = L(y).$$

In order to continue, let us assume that $C(y, \alpha) = A(\alpha)\, Y(y)$ and that (9) may be rewritten in the form

$$\frac{\partial}{\partial y} g(y, \alpha) = A(\alpha)\, Y(y)\, g(y, \alpha+1). \tag{10}$$

We now seek a transformation of $g(y, \alpha)$ which will provide factors $\dfrac{1}{A(\alpha)}$ and $\dfrac{1}{Y(y)}$. If we let $z = \int_{y_1}^{y} Y(v)\, dv$, where y_1 is a constant chosen to simplify the form of $F(z,\alpha)$, then

$$\frac{dz}{dy}=Y(y) \quad \text{and} \quad \frac{dy}{dz}=\frac{1}{Y(y)}. \tag{11}$$

Also we need a function $h(\alpha)$ which satisfies the identity

$$h(\alpha)=\frac{1}{A(\alpha)}h(\alpha+1).$$

The function $\exp\left\{\overset{\alpha}{\underset{\alpha_0}{S}}\log A(v)\,\Delta v\right\}$ is such a function. From the calculus of finite differences we have

$$\underset{\alpha}{\Delta}\overset{\alpha}{\underset{\alpha_0}{S}}\log A(v)\,\Delta v=\log A(\alpha).$$

Using this relation we are able to establish the following identity:

$$\begin{aligned}\exp\left\{\overset{\alpha}{\underset{\alpha_0}{S}}\log A(v)\,\Delta v\right\}&=\exp\left\{\overset{\alpha+1}{\underset{\alpha_0}{S}}\log A(v)\,\Delta v-\underset{\alpha}{\Delta}\overset{\alpha}{\underset{\alpha_0}{S}}\log A(v)\,\Delta v\right\}\\&=\exp\left\{\overset{\alpha+1}{\underset{\alpha_0}{S}}\log A(v)\,\Delta v\right\}\frac{1}{A(\alpha)}.\end{aligned} \tag{12}$$

Therefore, if we let

$$F(z,\alpha)=\exp\left\{\overset{\alpha}{\underset{\alpha_0}{S}}\log A(v)\,\Delta v\right\}g(y,\alpha), \tag{13}$$

we get

$$\frac{\partial}{\partial z}F(z,\alpha)=\exp\left\{\overset{\alpha}{\underset{\alpha_0}{S}}\log A(v)\,\Delta v\right\}[A(\alpha)\,Y(y)\,g(y,\alpha+1)]\frac{dy}{dz}. \tag{14}$$

By means of (11) and (12) we are able to simplify (14) as follows:

$$\begin{aligned}\frac{\partial}{\partial z}F(z,\alpha)&=\exp\left\{\overset{\alpha+1}{\underset{\alpha_0}{S}}\log A(v)\,\Delta v\right\}g(y,\alpha+1)\\&=F(z,\alpha+1).\end{aligned}$$

Since in the two transformations, $f(y,\alpha)$ to $g(y,\alpha)$ and $g(y,\alpha)$ to $F(z,\alpha)$, arbitrary constants y_0, y_1, and α_0 have been introduced, it is not evident that for a given f-type equation the resulting function $F(z,\alpha)$ is unique. Truesdell [1; p. 31, Theorem (7.2)] proved the following *uniqueness theorem:*

Suppose a function $f(y,\alpha)$ satisfies an equation of the type

$$\frac{\partial}{\partial y}f(y,\alpha)=A(y,\alpha)f(y,\alpha)+B(y,\alpha)f(y,\alpha+1),$$

which is reducible to the F-equation

$$\frac{\partial}{\partial z} F(z,\alpha) = F(z,\alpha+1).$$

Furthermore, suppose that by using only transformations of the type

$$g(y,\alpha) = M(y,\alpha)\, f\big(h(y),\alpha\big),$$

there are obtained from the function $f(y,\alpha)$ *two transformed functions* $F_1(z,\alpha)$ *and* $F_2(z,\alpha)$ *which both satisfy the F-equation. Then there exist a periodic function* $\pi(\alpha)$ *of period* 1 *and constants b and c such that*

$$F_2(z,\alpha) = \pi(\alpha)\, b^{\alpha}\, F_1(b\,z+c,\alpha).$$

By means of this uniqueness theorem and the generating function theorem which follows, we will see that it is not possible to obtain more than one essentially unique generating function from a given f-type equation. We will now state and prove *Truesdell's generating function theorem* (see Truesdell [1; p. 82, Theorem 14.1]):

If the function $F(z,\alpha)$ *satisfies the F-equation and if* $F(z+y,\alpha)$ *possesses a Taylor series, then this series may be put into the form*

$$F(z+y,\alpha) = \sum_{n=0}^{\infty} \frac{y^n}{n!} F(z,\alpha+n).$$

Proof. If $F(z+y,\alpha)$ possesses a Taylor's expansion about $y=0$, then

$$F(z+y,\alpha) = \sum_{n=0}^{\infty} \frac{\partial^n}{\partial y^n} F(z+y,\alpha)\bigg|_{y=0} \frac{y^n}{n!}.$$

By hypothesis $F(z,\alpha)$ satisfies the F-equation. Then

$$\frac{\partial}{\partial z} F(z+y,\alpha) = F(z+y,\alpha+1).$$

It follows by induction that

$$\frac{\partial^n}{\partial z^n} F(z+y,\alpha) = F(z+y,\alpha+n).$$

The coefficients of the expansion may then be determined as follows:

$$\begin{aligned}\frac{\partial^n}{\partial y^n} F(z+y,\alpha)\bigg|_{y=0} &= \frac{\partial^n}{\partial z^n} F(z+y,\alpha)\bigg|_{y=0} \\ &= F(z+y,\alpha+n)|_{y=0} = F(z,\alpha+n).\end{aligned}$$

Therefore

$$F(z+y,\alpha) = \sum_{n=0}^{\infty} F(z,\alpha+n)\frac{y^n}{n!}.$$

3. The Hermite polynomials $\{H_{\alpha+n}(x)\}$. We will show in detail how Truesdell's F-equation can be used to find a generating function for the set of Hermite polynomials $\{H_{\alpha+n}(x)\}$. Let $\phi_n(x)=H_n(x)/n!$. Then $\phi_n(x)$ satisfies

$$\phi_n'(x)=2x\,\phi_n(x)-(n+1)\,\phi_{n+1}(x). \tag{1}$$

By letting

$$f(y,\alpha)=\phi_\alpha(y)=\frac{H_\alpha(y)}{\Gamma(\alpha+1)},$$

we get from (1) the partial differential recurrence relation

$$\frac{\partial}{\partial y}f(y,\alpha)=2y\,f(y,\alpha)-(\alpha+1)\,f(y,\alpha+1). \tag{2}$$

Equation (2) corresponds to the equation which we have called the f-type equation,

$$\frac{\partial}{\partial y}f(y,\alpha)=A(y,\alpha)\,f(y,\alpha)+B(y,\alpha)\,f(y,\alpha+1),$$

with $A(y,\alpha)=2y$ and $B(y,\alpha)=-(\alpha+1)$.

We want to transform $f(y,\alpha)$ into $g(y,\alpha)$ so that

$$\frac{\partial}{\partial y}g(y,\alpha)=C(y,\alpha)\,g(y,\alpha+1).$$

Therefore, we let

$$\begin{aligned} g(y,\alpha)&=f(y,\alpha)\exp\left\{-\int_{y_0}^{y}A(v,\alpha)\,dv\right\}\\ &=f(y,\alpha)\exp\left\{-\int_{y_0}^{y}2v\,dv\right\}. \end{aligned}$$

By choosing $y_0=0$, we get

$$g(y,\alpha)=f(y,\alpha)\exp\{-y^2\}. \tag{3}$$

Then

$$\begin{aligned} \frac{\partial}{\partial y}g(y,\alpha)&=-(\alpha+1)\,f(y,\alpha+1)\exp\{-y^2\}\\ &=-(\alpha+1)\,g(y,\alpha+1). \end{aligned}$$

We have thus obtained the g-type equation

$$\frac{\partial}{\partial y}g(y,\alpha)=-(\alpha+1)\,g(y,\alpha+1). \tag{4}$$

Let $C(y,\alpha)$ denote the factorable coefficient of $g(y,\alpha+1)$ in equation (4). Then

$$C(y,\alpha)=(-1)(\alpha+1),$$

with $Y(y)=-1$ and $A(\alpha)=\alpha+1$. We effect the transformation of $g(y,\alpha)$ into $F(z,\alpha)$ by letting

$$z=\int_{y_1}^{y} Y(v)\,dv=\int_{y_1}^{y}(-1)\,dv=-y+y_1$$

and

$$\begin{aligned}F_0\,F(z,\alpha)&=g(y,\alpha)\exp\left\{\mathop{S}_{\alpha_0}^{\alpha}\log A(v)\,\Delta v\right\}\\&=g(y,\alpha)\exp\left\{\mathop{S}_{\alpha_0}^{\alpha}\log(1+v)\,\Delta v\right\}.\end{aligned}$$

If we choose $y_1=0$, $\alpha_0=-1$, and $F_0=1/(2\pi)^{\frac{1}{2}}$, we get

$$F(z,\alpha)=\Gamma(\alpha+1)\,g(-z,\alpha). \tag{5}$$

To determine the factor $\Gamma(\alpha+1)$ we have used the relation

$$\log\Gamma(x)=\mathop{S}_{0}^{x}\log z\,\Delta z+\log\sqrt{2\pi},$$

with $z=1+v$. See Milne-Thomson [1; p. 253]. However, since

$$(\alpha+1)\,\Gamma(\alpha+1)=\Gamma(\alpha+2),$$

we could have determined this particular transformation by inspection. Also instead of using $z=-y$ we could have used the fact that

$$(-1)\,e^{i\pi\alpha}=e^{i\pi(\alpha+1)}$$

and let

$$F(z,\alpha)=e^{i\pi\alpha}\,\Gamma(\alpha+1)\,g(z,\alpha),$$

with $z=y$.

To show that $F(z,\alpha)$ does indeed satisfy the F-equation, we determine $\frac{\partial}{\partial z}F(z,\alpha)$ as follows:

$$\begin{aligned}\frac{\partial}{\partial z}F(z,\alpha)&=\Gamma(\alpha+1)\,[-(\alpha+1)\,g(-z,\alpha+1)]\,(-1)\\&=\Gamma(\alpha+2)\,g(-z,\alpha+1)\\&=F(z,\alpha+1).\end{aligned}$$

For later use we will express $F(z,\alpha)$ in the following form:

$$\begin{aligned}F(z,\alpha)&=\Gamma(\alpha+1)\,f(-z,\alpha)\exp\{-z^2\}\\&=H_\alpha(-z)\exp\{-z^2\}.\end{aligned}$$

Were are now in a position to use Truesdell's generating function theorem [1; p. 82, Theorem (14.1)], proved in the preceding section. Since

$$F(z+y,\alpha)=\exp\{-(z+y)^2\}\,H_\alpha(-z-y)$$

and

$$F(z,\alpha+n)=\exp\{-z^2\}\,H_{\alpha+n}(-z),$$

we apply Truesdell's generating function theorem to get

$$\exp\{-(z+y)^2\}\,H_\alpha(-z-y)=\sum_{n=0}^{\infty}\frac{y^n}{n!}\exp\{-z^2\}\,H_{\alpha+n}(-z). \tag{6}$$

If we divide both members of (6) by $\exp\{-z^2\}$ and replace $-z$ by x, we get the generating relation

$$\exp\{2xy-y^2\}\,H_\alpha(x-y)=\sum_{n=0}^{\infty}H_{\alpha+n}(x)\frac{y^n}{n!}. \tag{7}$$

See Rainville [1; p. 197, (1)], Truesdell [1; p. 85, (10)], and Weisner [2; p. 144, (4.2)].

4. The descending equation. Suppose $\phi_n(x)$ is a given function satisfying the descending equation

$$\phi_n'(x)=A(x,n)\,\phi_n(x)+B(x,n)\,\phi_{n-1}(x), \tag{1}$$

where $B(x,n)\neq 0$. Let $f(y,\alpha)=\phi_\alpha(y)$. Then equation (1) implies

$$\frac{\partial}{\partial y}f(y,\alpha)=A(y,\alpha)\,f(y,\alpha)+B(y,\alpha)\,f(y,\alpha-1). \tag{2}$$

We seek to transform $f(y,\alpha)$ into a function $G(z,\alpha)$ which satisfies the equation

$$\frac{\partial}{\partial z}G(z,\alpha)=G(z,\alpha-1).$$

We will call this equation the G-equation. The procedure we use for effecting this transformation is patterned after the method Truesdell developed for the ascending case. For a function $\phi_\alpha(x)$ defined for negative α Truesdell [1; p. 83, (3)] used an adaptation of the method of Section 2.

As in the ascending case we let

$$g(y,\alpha)=f(y,\alpha)\exp\left\{-\int_{y_0}^{y}A(v,\alpha)\,dv\right\}.$$

The constant y_0 must be chosen so that the integral exists; otherwise y_0 is arbitrary. The partial derivative with respect to y of the transformed function $g(y,\alpha)$ will contain only one term, and that term can be expressed with $g(y,\alpha-1)$ as a factor:

$$\frac{\partial}{\partial y} g(y,\alpha)=B(y,\alpha)f(y,\alpha-1)\exp\left\{-\int_{y_0}^{y} A(v,\alpha)\,dv\right\}$$

$$=B(y,\alpha)\exp\left\{-\int_{y_0}^{y} \underset{\alpha}{\Delta} A(v,\alpha-1)\,dv\right\} g(y,\alpha-1).$$

THEOREM 1. *Let*

$$C(y,\alpha)=B(y,\alpha)\exp\left\{-\int_{y_0}^{y} \underset{\alpha}{\Delta} A(v,\alpha-1)\,dv\right\}$$

and

$$E(y,\alpha)=\frac{\partial}{\partial y}\log B(y,\alpha)-\underset{\alpha}{\Delta} A(y,\alpha-1).$$

The function $C(y,\alpha)$ *will satisfy the factorability condition,*

$$C(y,\alpha)=A(\alpha)\,Y(y),$$

if and only if $E(y,\alpha)$ *is a function of* y *independent of* α.

Proof. The function $C(y,\alpha)$ may be written as follows:

$$C(y,\alpha)=\exp\left\{\log B(y,\alpha)-\int_{y_0}^{y} \underset{\alpha}{\Delta} A(v,\alpha-1)\,dv\right\}.$$

Then

$$\frac{\partial}{\partial y} C(y,\alpha)/C(y,\alpha)=E(y,\alpha).$$

First, suppose that

$$C(y,\alpha)=A(\alpha)\,Y(y).$$

From this assumption it follows that

$$\frac{\partial}{\partial y} C(y,\alpha)/C(y,\alpha)=Y'(y)/Y(y).$$

Therefore, $E(y,\alpha)=Y'(y)/Y(y)$. We may now conclude that $E(y,\alpha)$ is dependent only on y if $C(y,\alpha)$ satisfies the factorability condition.

Next let us assume that

$$E(y,\alpha)=L(y),$$

where $L(y)$ is independent of α. It then follows that

$$\frac{\partial}{\partial y} C(y,\alpha)/C(y,\alpha)=L(y).$$

Accordingly we have

$$\log C(y,\alpha)=G(y)+H(\alpha)$$

and

$$C(y,\alpha)=e^{G(y)}\,e^{H(\alpha)},$$

where $G(y)$ is a function such that $G'(y)=L(y)$ and $H(\alpha)$ is introduced as an arbitrary function independent of y. (However, since $C(y,\alpha)$ is assumed to be a known function, $H(\alpha)$ may be determined.) We have thus proved that $C(y,\alpha)$ is factorable if $E(y,\alpha)=L(y)$.

We now seek a function $h(\alpha)$ which may be written in the form

$$h(\alpha)=\frac{1}{A(\alpha)}\,h(\alpha-1)$$

and a function $z(y)$ such that

$$\frac{dy}{dz}=\frac{1}{Y(y)}.$$

We would then have

$$G(z,\alpha)=h(\alpha)\,g(y,\alpha)$$

and

$$\begin{aligned}\frac{\partial}{\partial z}G(z,\alpha)&=\left[\frac{1}{A(\alpha)}\,h(\alpha-1)\right][A(\alpha)\,Y(y)\,g(y,\alpha-1)]\,\frac{dy}{dz}\\&=h(\alpha-1)\,g(y,\alpha-1)\\&=G(z,\alpha-1).\end{aligned}$$

We now prove that the function $\exp\left\{-\overset{\alpha+1}{\underset{\alpha_0}{S}}\log A(v)\,\Delta v\right\}$ from the calculus of finite differences satisfies the condition imposed on $h(\alpha)$. Using the identity

$$\exp\left\{-\overset{\alpha+1}{\underset{\alpha_0}{S}}\log A(v)\,\Delta v\right\}=\exp\left\{-\overset{\alpha}{\underset{\alpha_0}{S}}\log A(v)\,\Delta v-\underset{\alpha}{\Delta}\overset{\alpha}{\underset{\alpha_0}{S}}\log A(v)\,\Delta v\right\}$$

and the fact that

$$\underset{\alpha}{\Delta}\overset{\alpha}{\underset{\alpha_0}{S}}\log A(v)\,\Delta v=\log A(\alpha),$$

we find that

$$\exp\left\{-\overset{\alpha+1}{\underset{\alpha_0}{S}}\log A(v)\,\Delta v\right\}=\frac{1}{A(\alpha)}\exp\left\{-\overset{\alpha}{\underset{\alpha_0}{S}}\log A(v)\,\Delta v\right\}.$$

Accordingly, let

$$h(\alpha)=\exp\left\{-\overset{\alpha+1}{\underset{\alpha_0}{S}}\log A(v)\,\Delta v\right\}.$$

Also the condition that

$$\frac{dy}{dz}=\frac{1}{Y(y)}$$

will be satisfied if we let

$$z=\int_{y_1}^{y} Y(v)\,dv,$$

where y_1 is an arbitrary constant restricted only to guarantee the existence of the integral.

We may now state a generating function theorem for the descending case.

THEOREM 2. If the function $G(z,\alpha)$ satisfies the G-equation,

$$\frac{\partial}{\partial z}G(z,\alpha)=G(z,\alpha-1),$$

and if $G(z+y,\alpha)$ possesses a Taylor's series in powers of y, then this series may be expressed in the following form:

$$G(z+y,\alpha)=\sum_{n=0}^{\infty}\frac{y^n}{n!}G(z,\alpha-n).$$

The proof follows from Taylor's theorem and the fact that

$$\frac{\partial^n}{\partial y^n}G(z+y,\alpha)\bigg|_{y=0}=G(z+y,\alpha-n)\bigg|_{y=0}=G(z,\alpha-n).$$

5. The Hermite polynomials $\{H_{\alpha-n}(x)\}$. We will use the Hermite polynomials to illustrate the use the descending generating function theorem of the preceding section. The Hermite polynomial $H_n(x)$ satisfies the descending equation

$$H_n'(x)=2nH_{n-1}(x). \tag{1}$$

Let $f(y,\alpha)=H_\alpha(y)$. Then we may write (1) in the following form:

$$\frac{\partial}{\partial y}f(y,\alpha)=2\alpha f(y,\alpha-1). \tag{2}$$

By means of a single transformation determined by inspection we now transform $f(y,\alpha)$ into a function $G(z,\alpha)$ satisfying the G-equation. Let $z=y$ and let

$$G(z,\alpha)=\frac{1}{2^\alpha\Gamma(\alpha+1)}H_\alpha(z).$$

Then

$$\begin{aligned}\frac{\partial}{\partial z}G(z,\alpha)&=\frac{1}{2^{\alpha-1}\Gamma(\alpha)}H_{\alpha-1}(z)\\&=G(z,\alpha-1).\end{aligned}$$

Applying the generating function theorem of the preceding section, we get for the Hermite polynomials

$$\frac{H_\alpha(z+y)}{2^\alpha\Gamma(\alpha+1)}=\sum_{n=0}^{\infty}\frac{y^n}{n!}\,\frac{H_{\alpha-n}(z)}{2^{\alpha-n}\Gamma(\alpha-n+1)}$$

or equivalently

$$H_\alpha(z+y)=\sum_{n=0}^{\infty}\frac{(-2y)^n}{n!}(-\alpha)_n H_{\alpha-n}(z). \tag{3}$$

See Weisner [2; p. 144, (4.1)].

If α is an integer, the generating function of (3) has an expansion containing only a finite number of terms. This generating relation may be interpreted as an addition theorem for the Hermite polynomials. Also from (3) we may obtain a multiplication theorem by means of two simple substitutions. If in (3) we replace y by $y-z$, we get

$$H_\alpha(y)=\sum_{n=0}^{\infty}\frac{(-2)^n}{n!}(y-z)^n(-\alpha)_n H_{\alpha-n}(z). \tag{4}$$

If in (4) we replace y by kz, we get the following multiplication theorem:

$$H_\alpha(kz)=\sum_{n=0}^{\infty}\frac{(-2)^n}{n!}(k-1)^n z^n(-\alpha)_n H_{\alpha-n}(z). \tag{5}$$

See Truesdell [2] in which he points out that every generating expansion for an F-function may be interpreted as an addition or multiplication theorem.

6. The Charlier polynomials. We will obtain two generating functions for the Charlier polynomials $\{c_n(a;x)\}$ by taking advantage of their relation to the modified Laguerre polynomials $\{(-1)^n L_n^{a-n}(x)\}$. In harmony with Erdélyi [2; p. 226, (4)] we define the Charlier polynomial as follows:

$$\begin{aligned} c_n(x;a)&=\sum_{k=0}^{n}\binom{n}{k}\binom{x}{k}(-1)^k k!\,a^{-k}\\ &={}_2F_0\left(-n,\ -x;\ -;\ -\frac{1}{a}\right), \end{aligned}$$

where $a>0$ and $x=0,1,2,\ldots$. Since

$$\begin{aligned} f_n^{-a}(x)&=(-1)^n L_n^{a-n}(x)\\ &=\frac{x^n}{n!}\,{}_2F_0\left(-n,\ -a;\ -;\ -\frac{1}{x}\right), \end{aligned}$$

we may write

$$f_n^{-a}(x)=\frac{x^n}{n!}\,c_n(a;x), \tag{1}$$

where $x>0$ and $a=0, 1, 2, \ldots$. Therefore, generating functions for the set $\{f_n^{-a}(x)\}$ can be transformed into generating functions for the Charlier polynomials $\{c_n(a;x)\}$.

The functions $f_n^b(x)$ satisfies the following independent differential-difference equations:

$$x\,D f_n^b(x)=(x+n+b)\,f_n^b(x)-(n+1)\,f_{n+1}^b(x) \tag{2}$$

and

$$D f_n^b(x)=f_{n-1}^b(x), \quad \text{where} \quad D=\frac{d}{dx}. \tag{3}$$

If we let $f(y,\alpha)=f_\alpha^b(y)$, then Eq. (2) implies

$$\frac{\partial}{\partial y} f(y,\alpha)=\left(1+\frac{\alpha+b}{y}\right) f(y,\alpha)-\frac{\alpha+1}{y} f(y,\alpha+1).$$

We will now transform $f(y,\alpha)$ into a function $g(y,\alpha)$ such that

$$\frac{\partial}{\partial y} g(y,\alpha)=C(y,\alpha)\,g(y,\alpha+1).$$

Since for $y_0=1$

$$\exp\left\{-\int_{y_0}^{y}\left(1+\frac{\alpha+b}{v}\right) dv\right\}=e(e^{-y}\,y^{-\alpha-b}),$$

we let

$$g(y,\alpha)=e^{-y}\,y^{-\alpha-b} f(y,\alpha).$$

Then

$$\frac{\partial}{\partial y} g(y,\alpha)=-(\alpha+1)\,g(y,\alpha+1).$$

We observe that

$$[e^{i\pi\alpha}\,\Gamma(\alpha+1)]\,[(-1)(\alpha+1)]=e^{i\pi(\alpha+1)}\,\Gamma(\alpha+2).$$

Accordingly we make the second transformation by inspection. If we let $y=z$, we then have

$$F(z,\alpha)=e^{i\pi\alpha}\,\Gamma(\alpha+1)\,g(z,\alpha)$$

and

$$\frac{\partial F(z,\alpha)}{\partial z}=F(z,\alpha+1).$$

Using Truesdell's ascending generating function theorem, we get

$$e^{i\pi\alpha}\,\Gamma(\alpha+1)\,e^{-(z+y)}(z+y)^{-\alpha-b}f_\alpha^b(z+y)$$
$$=\sum_{n=0}^{\infty}\frac{y^n}{n!}\,e^{i\pi(\alpha+n)}\,\Gamma(\alpha+n+1)\,e^{-z}\,z^{-\alpha-n-b}f_{\alpha+n}^b(z),$$

or equivalently

$$e^{-y}\left(1+\frac{y}{z}\right)^{-b}[\Gamma(\alpha+1)(z+y)^{-\alpha}f_\alpha^b(z+y)]$$
$$=\sum_{n=0}^{\infty}\frac{(-y)^n}{n!}[\Gamma(\alpha+n+1)\,z^{-\alpha-n}f_{\alpha+n}^b(z)]. \tag{4}$$

The generating relation (4) is easily expressed in terms of Charlier polynomials:

$$e^{-y}\left(1+\frac{y}{z}\right)^{-b}c_\alpha(-b;z+y)=\sum_{n=0}^{\infty}\frac{(-y)^n}{n!}c_{\alpha+n}(-b;z).$$

Replacing z by x, y by $-y$, and b by $-b$, we get

$$e^{y}\left(1-\frac{y}{x}\right)^{b}c_\alpha(\mathrm{b};x-y)=\sum_{n=0}^{\infty}\frac{y^n}{n!}c_{\alpha+n}(b;x), \tag{5}$$

where $x>0$ and $b=0, 1, 2, \ldots$.

The generating relation of (5) is equivalent to Doetsch's formula given by Truesdell [1; p. 88, (26)], provided b is a positive integer:

$$\psi_\alpha(b,z+y)=\sum_{n=0}^{\infty}\frac{y^n}{n!}\psi_{\alpha+n}(b,z),$$

where

$$\psi_n(b,z)=\frac{z^b\,e^{-z}}{b!}\,p_n(b,z)$$

and

$$p_n(b,z)=(-1)^n\,c_n(b,z).$$

We now turn to the problem of generating the set $\{c_{\alpha-n}(a;x)\}$ with α fixed. Since

$$\frac{d}{dx}f_n^{-a}(x)=f_{n-1}^{-a}(x),$$

we let

$$G(z,\alpha)=f_\alpha^{-a}(z).$$

Then

$$\frac{\partial}{\partial z}G(z,\alpha)=f_{\alpha-1}^{-a}(z)=G(z,\alpha-1).$$

Therefore, we may apply the descending generating function theorem of Section 4:

$$f_\alpha^{-a}(z+y)=\sum_{n=0}^{\infty}\frac{y^n}{n!}f_{\alpha-n}^{-a}(z).$$

If we express this generating relation in terms of Charlier polynomials, we get

$$\frac{(z+y)^\alpha}{\Gamma(\alpha+1)}c_\alpha(a;z+y)=\sum_{n=0}^{\infty}\frac{y^n}{n!}\frac{z^{\alpha-n}}{\Gamma(\alpha-n+1)}c_{\alpha-n}(a;z).$$

Using the fact that

$$\frac{\Gamma(\alpha+1)}{\Gamma(\alpha+1-n)}=(-1)^n(-\alpha)_n$$

and letting $t=-\dfrac{y}{z}$, we get

$$(1-t)^\alpha c_\alpha(a;z-zt)=\sum_{n=0}^{\infty}\frac{t^n}{n!}(-\alpha)_n c_{\alpha-n}(a;z).$$

Returning to the notation used in Erdélyi [2; p. 226] in which x denotes the discrete variable, we rewrite the two generating functions obtained in this section as follows:

For $a>0$ and $x=0,1,2,\ldots$,

$$e^t\left(1-\frac{t}{a}\right)^x c_\alpha(x;a-t)=\sum_{n=0}^{\infty}\frac{t^n}{n!}c_{\alpha+n}(x;a)$$

and

$$(1-t)^\alpha c_\alpha(x;a(1-t))=\sum_{n=0}^{\infty}\frac{t^n}{n!}(-\alpha)_n c_{\alpha-n}(x;a).$$

Also see Truesdell [3; p. 453, (23) and (26)] for equivalent generating functions in more concise form.

Chapter V

Miscellaneous Methods

1. Introduction. Most of the methods described in this book have been developed in the past twenty years. Truesdell's F-equation method was published in 1948. Weisner's first paper on his group theoretic method was published in 1955. Some of the direct summation techniques developed by Rainville were published in 1960.

Although it is the primary purpose of this study to bring to the reader's attention these three widely applicable methods, we include in this chapter some other useful methods.

2. Classes of generating functions. In accordance with Boas and Buck terminology, a set of polynomials has a generalized Appell representation if it is generated by the formal relation

$$A(t)\, C\big(x\, B(t)\big) = \sum_{n=0}^{\infty} \phi_n(x)\, t^n, \tag{1}$$

where

$$A(t) = \sum_{n=0}^{\infty} a_n\, t^n, \qquad a_0 \neq 0, \tag{2}$$

$$B(t) = \sum_{n=1}^{\infty} b_n\, t^n, \qquad b_1 \neq 0, \tag{3}$$

and

$$C(t) = \sum_{n=0}^{\infty} c_n\, t^n, \qquad c_n \neq 0 \text{ for all } n. \tag{4}$$

See Boas and Buck [1; p. 18].

Among the sets included in the Boas and Buck classification are the Brenke polynomials, which satisfy a generating relation of the type

$$A(t)\, C(x\, t) = \sum_{n=0}^{\infty} \phi_n(x)\, t^n, \tag{5}$$

the Sheffer A-type zero polynomials which satisfy

$$A(t) \exp\{x\, B(t)\} = \sum_{n=0}^{\infty} \phi_n(x)\, t^n, \tag{6}$$

and the Appell polynomials which satisfy

$$A(t)\exp\{x\,t\}=\sum_{n=0}^{\infty}\phi_n(x)\,t^n. \tag{7}$$

See Boas and Buck [2; p. 626].

The sets of Sheffer A-type zero polynomials constitute a subclass of the class of Sheffer A-type m polynomials, where m is a nonnegative integer. We now define this class.

Let

$$J(x,D)=\sum_{k=0}^{\infty}p_k(x)\,D^{k+1},$$

where $D=d/dx$ and $p_k(x)$ is a polynomial whose degree does not exceed k. For any simple set of polynomials $\{\phi_n(x)\}$ there exists a unique operator $J(x,D)$ such that

$$J(x,D)\,\phi_n(x)=\phi_{n-1}(x), \qquad n\geqq 1.$$

(See Sheffer [1; p. 592, Theorem 1.1] or Rainville [1; p. 218, Theorem 70].) The polynomial set $\{\phi_n(x)\}$ is said to belong to this operator. If the maximum degree of any element of the set $\{p_k(x)\}$ is m, then the set $\{\phi_n(x)\}$ belonging to the operator $J(x,D)$ is said to be of Sheffer A-type m.

As an example, let $\phi_n(x)=H_n(x)/n!$, where $\{H_n(x)\}$ is the set of Hermite polynomials. Then the operator associated with this set is $\frac{1}{2}D$ since

$$\frac{1}{2}D\left[\frac{H_n(x)}{n!}\right]=\frac{1}{2}\left[\frac{2n\,H_{n-1}(x)}{n!}\right]=\frac{H_{n-1}(x)}{(n-1)!}.$$

Therefore, the set $\{H_n(x)/n!\}$ is said to be of Sheffer A-type zero.

Also consider the sets $\{L_n^a(x)\}$ and $\{L_n^a(x)/(a+1)_n\}$, where $L_n^a(x)$ is the Laguerre polynomial. It can be proved that

$$D(D-1)^{-1}L_n^a(x)=L_{n-1}^a(x)$$

and

$$[x\,D^2+(a+1)\,D]\frac{L_n^a(x)}{(a+1)_n}=\frac{L_{n-1}^a(x)}{(a+1)_{n-1}}.$$

Therefore, the set $\{L_n^a(x)\}$ is of Sheffer A-type zero, and the set $\{L_n^a(x)/(a+1)_n\}$ is of Sheffer A-type one.

To establish the equivalence of the two definitions which we have given for the class of Sheffer A-type zero polynomials, we quote the following theorem from Rainville [1; p. 222]:

THEOREM A. *A necessary and sufficient condition that $\phi_n(x)$ be of Sheffer A-type zero (by operator definition) is that $\phi_n(x)$ possess the* *generating function indicated in*

$$A(t)\exp\{xH(t)\}=\sum_{n=0}^{\infty}\phi_n(x)\,t^n,$$

in which $H(t)$ and $A(t)$ have (at least the formal) expansions

$$H(t)=\sum_{n=0}^{\infty}h_n t^{n+1}, \qquad h_0\neq 0,$$

$$A(t)=\sum_{n=0}^{\infty}a_n t^n, \qquad a_0\neq 0.$$

Let $J(t)$ denote the inverse of $H(t)$ of Theorem A, i.e.,

$$H(J(t))=J(H(t))=t,$$

then $J(D)$ is the operator associated with the set $\{\phi_n(x)\}$.

Of the various other theorems which furnish necessary and sufficient conditions for a set to belong to a certain class, we give only one.

THEOREM B. *A necessary and sufficient condition that $\{\phi_n(x)\}$ be an Appell set (by generating function definition) is that*

$$\phi_n'(x)=\phi_{n-1}(x), \qquad n\geqq 1.$$

We note that Theorem B is actually a corollary of Theorem A with $H(t)=J(t)=t$. Hence, every Appell set belongs to the operator D.

We call the reader's attention to the possibility of finding a generating function by first noting that a given set satisfies a necessary and sufficient condition for a certain classification. For example, suppose for some set $\{\phi_n(x)\}$ a differential operator $J(D)$ is found which is independent of x and which satisfies the relation

$$J(D)\,\phi_n(x)=\phi_{n-1}(x), \qquad n\geqq 1.$$

It then follows that the set $\{\phi_n(x)\}$ is of Sheffer A-type zero and possesses a generating function of the form

$$A(t)\exp\{xH(t)\}=\sum_{n=0}^{\infty}\phi_n(x)\,t^n,$$

where $H(t)$ is the inverse of $J(t)$. Only $A(t)$ would remain to be determined by summation methods. We illustrate this procedure for the special case for which $J(D)=D$, and the set is an Appell set.

Suppose that we want to find a generating function for the Bernoulli polynomials given that

$$B_0(x)=1, \tag{8}$$

$$DB_n(x)=nB_{n-1}(x), \tag{9}$$

and

$$\Delta B_n(x) = B_n(x+1) - B_n(x) = n\,x^{n-1}. \tag{10}$$

From (9) we know that $\{B_n(x)/n!\}$ is an Appell set. Then there exists a function $A(t)$ such that

$$e^{xt} A(t) = \sum_{n=0}^{\infty} \frac{B_n(x)}{n!} t^n. \tag{11}$$

Since $B_0(x+1) - B_0(x) = 1 - 1 = 0$, we sum from $n = 1$ and use (10) to write

$$\sum_{n=1}^{\infty} [B_n(x+1) - B_n(x)] \frac{t^n}{n!} = \sum_{n=1}^{\infty} [n\,x^{n-1}] \frac{t^n}{n!}.$$

Then

$$\sum_{n=1}^{\infty} [B_n(x+1) - B_n(x)] \frac{t^n}{n!} = \sum_{n=0}^{\infty} \frac{x^n\,t^{n+1}}{n!},$$

or equivalently,

$$e^{(x+1)t} A(t) - e^{xt} A(t) = t\,e^{xt}. \tag{12}$$

From (12) we find

$$A(t) = \frac{t}{e^t - 1}.$$

Therefore, the generating relation we seek is

$$e^{xt} \left(\frac{t}{e^t - 1} \right) = \sum_{n=0}^{\infty} \frac{B_n(x)}{n!} t^n.$$

See Jordan [1; p. 250 (1)].

Sometimes it is more convenient to work with "a reversed set" than with the original set. For any given set of polynomials $\{p_n(x)\}$ we may represent $p_n(x)$ as follows:

$$p_n(x) = a_0\,x^n + a_1\,x^{n-1} + \cdots + a_{n-1}\,x + a_n.$$

Then

$$x^n\,p_n\left(\frac{1}{x}\right) = a_0 + a_1\,x + \cdots + a_{n-1}\,x^{n-1} + a_n\,x^n.$$

It is customary to write

$$p_n^*(x) = x^n\,p_n\left(\frac{1}{x}\right).$$

The set $\{p_n^*(x)\}$ is called the reversed set with reference to the original set $\{p_n(x)\}$. If a generating relation can be found for either set, it can be easily transformed into one for the other set. Suppose that we are given the generating relation

$$G(x,t)=\sum_{n=0}^{\infty} p_n(x)\,t^n.$$

If we replace x by $1/x$ and t by $x\,t$, we get

$$\begin{aligned} G(1/x, x\,t) &= \sum_{n=0}^{\infty} p_n(1/x)(x\,t)^n \\ &= \sum_{n=0}^{\infty} p_n^*(x)\,t^n. \end{aligned}$$

Since for any Appell set $\{p_n(x)\}$ we have a generating relation of the type

$$A(t)\exp\{x\,t\}=\sum_{n=0}^{\infty} p_n(x)\,t^n,$$

where $A(u)=\sum_{n=0}^{\infty}\gamma_n u^n$ with $\gamma_0 \neq 0$, then it follows that a reversed set $\{p_n^*(x)\}$ of an Appell set must satisfy a relation of the type

$$A(x\,t)\exp\{t\}=\sum_{n=0}^{\infty} p_n^*(x)\,t^n.$$

Such sets are called simply *reversed Appell polynomials.*

We quote from Rainville [1; p. 134] a theorem which provides a second generating function for each reversed Appell set:

"From

$$e^t\,\psi(x\,t)=\sum_{n=0}^{\infty}\sigma_n(x)\,t^n, \qquad \psi(u)=\sum_{n=0}^{\infty}\gamma_n u^n,$$

it follows that for arbitrary c

$$(1-t)^{-c}F\left(\frac{x\,t}{1-t}\right)=\sum_{n=0}^{\infty}(c)_n\,\sigma_n(x)\,t^n,$$

in which

$$F(u)=\sum_{n=0}^{\infty}(c)_n\,\gamma_n u^n."$$

We now rewrite this theorem in terms of Appell polynomials:

THEOREM C. *From*

$$e^{xt}A(t)=\sum_{n=0}^{\infty} p_n(x)\,t^n, \qquad A(u)=\sum_{n=0}^{\infty}\gamma_n u^n,$$

it follows that for arbitrary c

$$(1-xt)^{-c} F\left(\frac{t}{1-xt}\right) = \sum_{n=0}^{\infty} (c)_n p_n(x) t^n,$$

in which

$$F(u) = \sum_{n=0}^{\infty} (c)_n \gamma_n u^n.$$

This transformed version of the original theorem permits us to write immediately a generating relation, containing an arbitrary parameter c, for any set whose Appell-type generating relation is given. For the set $\{f_n^a(x)\}$ satisfying

$$e^{xt}(1-t)^{-a} = \sum_{n=0}^{\infty} f_n^a(x) t^n,$$

we have

$$A(t) = (1-t)^{-a} = \sum_{n=0}^{\infty} \frac{(a)_n}{n!} t^n$$

and

$$F\left(\frac{t}{1-xt}\right) = \sum_{n=0}^{\infty} \frac{(c)_n (a)_n}{n!} \left(\frac{t}{1-xt}\right).$$

Therefore, the set $\{f_n^a(x)\}$ satisfies the (divergent) generating relation

$$(1-xt)^{-c} {}_2F_0\left[\begin{matrix} c, & a; \\ & -; \end{matrix} \; \frac{t}{1-xt}\right] \cong \sum_{n=0}^{\infty} (c)_n f_n^a(x) t^n.$$

See Chapter 1, Section 3, (3).

3. Natural pairs of generating functions. James Ward Brown [1; p. 34] made a study of generating functions occurring in "natural pairs" as described in the following theorem:

"*THEOREM* D. *Suppose that the modified Laguerre set* $\{L_n^{\alpha(n)}(x)\}$ *has a Sheffer A-type zero generating function*

$$A(t) \exp\big(x B(t)\big) = \sum_{n=0}^{\infty} c_n L_n^{\alpha(n)}(x) t^n. \tag{1}$$

If $\{\phi_n^a(x)\}$ *is a polynomial set generated by*

$$(1-t)^{-a-1} f(x,t) = \sum_{n=0}^{\infty} \phi_n^a(x) \frac{t^n}{n!}, \tag{2}$$

where $f(x,t)$ is independent of the parameter a, then

$$A(t)(1-B(t))^{a-1} f(x, B(t)) = \sum_{n=0}^{\infty} c_n (-1)^n \phi_n^{-a-\alpha(n)-n}(x) \frac{t^n}{n!} \tag{3}$$

and

$$A(t)(1-B(t))^a f\left(x, \frac{B(t)}{B(t)-1}\right) = \sum_{n=0}^{\infty} c_n \phi_n^{a+\alpha(n)}(x) \frac{t^n}{n!}.\text{"} \tag{4}$$

Brown's proof of this theorem is based on a symbolic generating function theorem which he had proved earlier. (See Brown [1; p. 5].)

If in the conclusion of the theorem quoted above we let $\alpha(n)=0$, we see that the sets $\{\phi_n^a(x)\}$ and $\{(-1^n \phi_n^{-a-n}(x)\}$ form a natural pair provided only that the set $\{\phi_n^a(x)\}$ can be generated by a function of the type $(1-t)^{-a-1} f(x,t)$, where $f(x,t)$ is independent of the parameter a. For example, in the theorem we replace (1) by

$$(1-t)^{-1} \exp\left\{\frac{-xt}{1-t}\right\} = \sum_{n=0}^{\infty} L_n(x)\, t^n \tag{5}$$

and (2) by

$$(1-t)^{-a-1}(1-xt)^{-1} = \sum_{n=0}^{\infty} g_n^a(x)\, t^n, \tag{6}$$

where $\{g_n^a(x)\}$ is the symbol arbitrarily used to represent the set of functions defined by (6). (See Erdélyi [3; p. 245, (2)].) In accordance with the notation of the theorem we then have

$$A(t)=(1-t)^{-1}, \quad B(t)=\frac{-t}{1-t}, \quad \text{and} \quad f(x,t)=(1-xt)^{-1}.$$

By substituting in (3) and simplifying we get

$$(1-t)^{-a+1}(1-t+xt)^{-1} = \sum_{n=0}^{\infty} (-1)^n g_n^{-a-n}(x)\, t^n. \tag{7}$$

(See Erdélyi [3; p. 247, (16), with $b=1$ and $c=a$].) The generating functions of (6) and (7) are, in Brown's terminology, a natural pair.

Generating functions of the type described in (2), i.e.,

$$(1-t)^{-a-1} f(x,t) = \sum_{n=0}^{\infty} \frac{\phi_n^a(x)}{n!} t^n.$$

have a special meaning which it seems appropriate to discuss at this time. Suppose $f(x,t)$ can be expanded in powers of t generating a set of functions $\{g_k(x)\}$:

$$f(x,t) = \sum_{k=0}^{\infty} g_k(x)\, t^k.$$

Then

$$(1-t)^{-1}f(x,t)=\sum_{n=0}^{\infty}t^n\sum_{k=0}^{\infty}g_k(x)\,t^k$$
$$=\sum_{n=0}^{\infty}\sum_{k=0}^{n}g_k(x)\,t^n.$$

Let $h_n(x)=\sum_{k=0}^{n}g_k(x)$. Then $h_n(x)$ is merely the sum of the finite set $\{g_0(x), g_1(x), \ldots, g_n(x)\}$. If $\{g_n(x)\}$ is a simple set of polynomials, then $\{h_n(x)\}$ is also a simple set of polynomials.

We now observe the effect of the factor $(1-t)^{-a-1}$, where a is a non-negative integer:

$$(1-t)^{-a-1}f(x,t)=\sum_{n=0}^{\infty}\frac{(a+1)_n}{n!}t^n\sum_{k=0}^{\infty}g_k(x)\,t^k$$
$$=\sum_{n=0}^{\infty}\sum_{k=0}^{n}\binom{a+n-k}{a}g_k(x)\,t^n.$$

The function

$$\frac{1}{\binom{a+n}{a}}\sum_{k=0}^{n}\binom{a+n-k}{a}g_k(x)$$

is called the a-th Cesàro mean of the finite set $\{g_0(x), g_1(x), \ldots, g_n(x)\}$. See Obrechkoff [1] with our a and k replaced by k and μ, respectively.

We now return to Brown's Theorem D on natural pairs. If in this theorem we replace (2) by

$$(1-t)^{-a-1}\exp\left\{\frac{-xt}{1-t}\right\}=\sum_{n=0}^{\infty}L_n^a(x)\,t^n, \tag{8}$$

we get the corollary specifically applicable to modified Laguerre sets.

"*COROLLARY* D-1. *Suppose that the modified Laguerre set* $\{L_{n(x)}^{\alpha(n)}\}$ *has a Sheffer A-type zero generating function,*

$$A(t)\exp(xB(t))=\sum_{n=0}^{\infty}c_nL_n^{\alpha(n)}(x)\,t^n.$$

Then

$$A(t)(1-B(t))^{a-1}\exp\left(x\frac{B(t)}{B(t)-1}\right)=\sum_{n=0}^{\infty}c_n(-1)^nL_n^{-a-\alpha(n)-n}(x)\,t^n$$

and

$$A(t)(1-B(t))^a\exp(xB(t))=\sum_{n=0}^{\infty}c_nL_n^{a+\alpha(n)}(x)\,t^n."$$

See Brown [1; p. 38].

If we choose $\alpha(n)=0$ and use

$$(1-t)^{-1}\exp\left\{\frac{-xt}{x-t}\right\}=\sum_{n=0}^{\infty} L_n(x)\,t^n$$

as the generating relation in the hypothesis of the corollary, we have $A(t)=(1-t)^{-1}$ and $B(t)=-t/(1-t)$. Then the paired generating relations are

$$(1-t)^{-a}\exp\{xt\}=\sum_{n=0}^{\infty}(-1)^n L_n^{-a-n}(x)\,t^n$$

and

$$(1-t)^{-1-a}\exp\left\{\frac{-xt}{1-t}\right\}=\sum_{n=0}^{\infty} L_n^a(x)\,t^n.$$

If $\alpha(n)=n$, we see that the paired sets are $\{(-1)^n L_n^{-a-2n}(x)\}$ and $\{L_n^{a+n}(x)\}$. These are the two sets in which Brown [1; pp. 32–40] was most interested. We will study these sets further in Section 4.

We now turn our attention Brown's method of finding pairs of generating functions by using an auxiliary parameter. Suppose that two given sets have a common parameter a. Suppose also that, by introducing a second parameter m, we are able to construct a set $\{p_n^a(m;x)\}$ such that the two given sets are special cases of the constructed set. Then if we are able to find a generating function for the constructed set $\{p_n^a(m;x)\}$, we will automatically find a pair of generating functions for the two given sets.

As a simple illustration of this method we choose the sets $\{L_n^a(x)\}$ and $\{(-1)^n L_n^{-a-n}(x)\}$. We assume as given

$$L_n^a(x)=\sum_{k=0}^{n}\frac{(a+1)_n(-1)^k}{(a+1)_k}\,\frac{x^k}{(n-k)!\,k!}$$

and

$$(-1)^n L_n^{-a-n}(x)=\sum_{k=0}^{n}(a)_{n-k}\frac{x^k}{(n-k)!\,k!}.$$

By examining these two expansions we are able to construct a generalization $p_n^a(m;x)$ in the following form:

$$p_n^a(m;x)=\sum_{k=0}^{n}(a+1+mk)_{n-k}(-1)^{mk}\frac{x^k}{(n-k)!\,k!},$$

where m is a nonnegative integer. Then

$$p_n^a(1;x)=L_n^a(x)$$

and

$$p_n^{a-1}(0;x)=(-1)^n L_n^{-a-n}(x).$$

We use direct summation techniques to obtain a generating function for the set $\{p_n^a(m; x)\}$:

$$\sum_{n=0}^{\infty} p_n^a(m; x)\, t^n = (1-t)^{-a-1} \exp\left\{\frac{x t}{(t-1)^m}\right\}.$$

Then as our pair of generating functions, we have

$$\sum_{n=0}^{\infty} L_n^a(x)\, t^n \equiv \sum_{n=0}^{\infty} p_n^a(1; x)\, t^n = (1-t)^{-a-1} \exp\left\{\frac{x t}{t-1}\right\}$$

and

$$\sum_{n=0}^{\infty} (-1)^n L_{n(x)}^{-a-n}\, t^n \equiv \sum_{n=0}^{\infty} p_n^{a-1}(0; x)\, t^n = (1-t)^{-a} \exp\{x t\}.$$

This pair is the same as the pair obtained from Brown's Corollary D-1 with $\alpha(n)=0$, but the auxiliary parameter method is entirely independent of Theorem D and its corollaries.

4. Generating functions in differentiated form or in integrated form. If in a given generating function $G(x, t)$ we consider x as fixed and t as a variable, we may change the form of the generating function by either differentiation or integration. For example, assume as given for the Legendre polynomials $\{P_n(x)\}$ the generating relation

$$(1-2x t+t^2)^{-1/2} = \sum_{n=0}^{\infty} P_n(x)\, t^n. \tag{1}$$

A differentiated form of this relation is

$$(1-2x t+t^2)^{-3/2}(1-t^2) = \sum_{n=0}^{\infty} (2n+1)\, P_n(x)\, t^n. \tag{2}$$

See Erdélyi [3; p. 246, (6)]. The procedure for obtaining (2) from (1) consists of replacing t by t^2 to introduce the $2n$,

$$(1-2x t^2+t^4)^{-1/2} = \sum_{n=0}^{\infty} P_n(x)\, t^{2n},$$

multiplying by t,

$$t(1-2x t^2+t^4)^{-1/2} = \sum_{n=0}^{\infty} P_n(x)\, t^{2n+1},$$

and differentiating with respect to t,

$$(1-2x t^2+t^4)^{-3/2}(1-t^4) = \sum_{n=0}^{\infty} P_n(x)(2n+1)\, t^{2n}.$$

If we now replace t^2 by t, we get

$$(1-2xt+t^2)^{-3/2}(1-t^2)=\sum_{n=0}^{\infty}(2n+1)P_n(x)\,t^n,$$

which is the desired differentiated form.

Such a differentiated form may be used to determine a generating function for a related set. For example, if we multiply both members of (2) by $(1-t)^{-a-1}$, we get

$$(1-t)^{-a-1}\frac{(1-t^2)}{(1-2xt+t^2)^{3/2}} = \sum_{n=0}^{\infty}\sum_{k=0}^{n}\frac{\Gamma(a+1+n-k)}{\Gamma(a+1)\,\Gamma(n-k+1)}(2k+1)P_k(x)\,t^n. \tag{3}$$

If we let

$$g_n(x)=\sum_{k=0}^{n}\frac{\Gamma(a+1+n-k)}{\Gamma(a+1)\,\Gamma(n-k+1)}(2k+1)P_k(x),$$

then we may rewrite (3) in the form

$$\frac{1+t}{(1-t)^a(1-2xt+t^2)^{3/2}}=\sum_{n=0}^{\infty}g_n(x)\,t^n.$$

See Erdélyi [3; p. 246, (8) and (9)].

Obtaining an integrated form of a given generating relation by actual integration is of course much more difficult. Suppose that for an arbitrary set $\{g_n^a(x)\}$ we are given the generating relation

$$G(x,t)=\sum_{n=0}^{\infty}g_n^a(x)\,t^n$$

and required to find $H(x,t)$ such that

$$H(x,t)=\sum_{n=0}^{\infty}\frac{a}{a+n}\,g_n^a(x)\,t^n.$$

Theoretically, we could proceed as follows:

$$t^{a-1}G(x,t)=\sum_{n=0}^{\infty}g_n^a(x)\,t^{n+a-1},$$

$$\int_0^t \tau^{a-1}G(x,\tau)\,d\tau=\sum_{n=0}^{\infty}g_n^a(x)\frac{t^{n+a}}{n+a}.$$

Then

$$a\,t^{-a}\int_0^t \tau^{a-1}G(x,\tau)\,d\tau=\sum_{n=0}^{\infty}\frac{a}{a+n}\,g_n^a(x)\,t^n.$$

The indicated integration may be too complicated to make such a procedure feasible. Therefore, we now consider two other methods for obtaining integrated forms.

As an illustration of the first of these methods, we assume as given for the Laguerre polynomials the generating relation

$$(1-t)^{-1-a}\exp\left\{\frac{-xt}{1-t}\right\}=\sum_{n=0}^{\infty}L_n^a(x)\,t^n \tag{4}$$

and seek a function $H(x,t)$ such that

$$H(x,t)=\sum_{n=0}^{\infty}\frac{a}{a+n}L_n^a(x)\,t^n.$$

We shall also assume as given the following generating relation

$$(1-t)^{-c}\,{}_1F_1\left[\begin{matrix}c; & \dfrac{-xt}{1-t}\\ a+1; & \end{matrix}\right]=\sum_{n=0}^{\infty}\frac{(c)_n L_n^a(x)\,t^n}{(a+1)_n}, \tag{5}$$

where c is any positive real number. The arbitrary constant is to be chosen so that a numerator parameter will be one unit less than a denominator parameter. If in (5) we choose $c=a$, we get

$$(1-t)^{-a}\,{}_1F_1\left[\begin{matrix}a; & \dfrac{-xt}{1-t}\\ a+1; & \end{matrix}\right]=\sum_{n=0}^{\infty}\frac{(a)_n L_n^a(x)\,t^n}{(a+1)_n},$$

or equivalently,

$$(1-t)^{-a}\,{}_1F_1\left[\begin{matrix}a; & \dfrac{-xt}{1-t}\\ a+1; & \end{matrix}\right]=\sum_{n=0}^{\infty}\frac{a}{a+n}L_n^a(x)\,t^n. \tag{6}$$

Then (6) is the desired integrated form of (4). Since both (4) and (6) are special cases of (5), we see that (5) provides both a generating relation and an integrated form of that relation.

For the second method of obtaining an integrated form of a given generating relation we return to Brown [1; pp. 55–62] and make illustrative use of the sets

$$\{(-1)^n L_n^{-a-2n}(x)\} \quad\text{and}\quad \{L_n^{a+n}(x)\}.$$

He studied these sets as a pair. We will use these sets (one at a time) to illustrate a series manipulation method for finding a generating relation and an integrated form of that relation.

For convenience let

$$g_n^a(x)=(-1)^n L_n^{-a-2n}(x).$$

Then

$$g_n^a(x)=\sum_{k=0}^{n}\frac{(a)_{2n-k}\,x^k}{(a)_n(n-k)!\,k!}.$$

From $g_n^a(x)$ we construct a new function $h_n^a(x)$ by *increasing the denominator parameter by one unit*:

$$h_n^a(x)=\sum_{k=0}^{n}\frac{(a)_{2n-k}\,x^k}{(a+1)_n(n-k)!\,k!}=\frac{a}{a+n}\,g_n^a(x).$$

We have thus constructed a function $h_n^a(x)$ which is an "integrated form" of $g_n^a(x)$. Such a construction is possible when the parameter a appears in a factorial function which is independent of k and also appears in at least one other factor of the given expansion of $g_n^a(x)$. We seek to determine $G(x,t)$ and $H(x,t)$ such that

$$G(x,t)=\sum_{n=0}^{\infty} g_n^a(x)\,t^n \tag{7}$$

and

$$H(x,t)=\sum_{n=0}^{\infty} h_n^a(x)\,t^n=\sum_{n=0}^{\infty}\frac{a}{a+n}\,g_n^a(x)\,t^n. \tag{8}$$

By using direct summation we find

$$G(x,t)=\frac{1}{\sqrt{1-4t}}\left(\frac{2}{1+\sqrt{1-4t}}\right)^{a-1}\exp\left\{\frac{2xt}{1+\sqrt{1-4t}}\right\}$$

and

$$H(x,t)=\left(\frac{2}{1+\sqrt{1-4t}}\right)^{a}{}_1F_1\left[\begin{matrix}a; \\ a+1;\end{matrix}\ \frac{2xt}{1+\sqrt{1-4t}}\right].$$

For the set $\{L_n^{a+n}(x)\}$ we use a similar procedure to find a generating relation and an integrated form of it. For the purposes of this explanation let

$$p_n^a(x)=L_n^{a+n}(x).$$

Then

$$p_n^a(x)=\sum_{k=0}^{n}\frac{(-1)^k(a+1)_{2n}\,x^k}{(a+1)_{n+k}(n-k)!\,k!}.$$

The parameter a appears in the factorial function $(a+1)_{2n}$, which is independent of k, and the parameter a also appears in one other factorial function. From $p_n^a(x)$ we construct a new function $q_n^a(x)$ by *decreasing the numerator parameter by one unit*:

$$q_n^a(x)=\sum_{k=0}^{n}\frac{(-1)^k(a)_{2n}\,x^k}{(a+1)_{n+k}(n-k)!\,k!}=\frac{a}{a+2n}\,p_n^a(x).$$

The generating relations for $\{p_n^a(x)\}$ and $\{q_n^a(x)\}$ are found to be

$$\sum_{n=0}^{\infty} p_n^a(x)\,t^n=\frac{1}{\sqrt{1-4t}}\left(\frac{2}{1+\sqrt{1-4t}}\right)^{a}\exp\left\{\frac{-4xt}{(1+\sqrt{1-4t})^2}\right\} \tag{9}$$

and

$$\sum_{n=0}^{\infty} q_n^a(x)\, t^n = \left(\frac{2}{1+\sqrt{1-4t}}\right)^a {}_1F_1\left[\begin{matrix}\frac{a}{2}; \\ \frac{a}{2}+1;\end{matrix} \quad \frac{-4xt}{(1+\sqrt{1-4t})^2}\right]. \tag{10}$$

The set $\{(-1)^n L_n^{-a-2n}(x)\}$ is a special case of the set of functions described in the following theorem:

THEOREM 1. *If*

$$\sum_{n=0}^{\infty} f_n(x)\, t^n = \frac{1}{\sqrt{1-4t}}\left(\frac{2}{1+\sqrt{1-4t}}\right)^{a-1} F\left(\frac{2xt}{1+\sqrt{1-4t}}\right)$$

where $F(u) = \sum_{k=0}^{\infty} \gamma_k u^k$, *then*

$$\sum_{n=0}^{\infty} \frac{a}{a+n} f_n(x)\, t^n = \left(\frac{2}{1+\sqrt{1-4t}}\right)^a \psi\left(\frac{2xt}{1+\sqrt{1-4t}}\right)$$

where

$$\psi(u) = \sum_{k=0}^{\infty} \frac{a}{a+k} \gamma_k x^k.$$

We will omit the proof of Theorem 1, since its proof parallels the proof (given below) of Theorem 2.

The set $\{L_n^{a+n}(x)\}$ is a special case of the set of functions described in the following theorem:

THEOREM 2. *If*

$$\sum_{n=0}^{\infty} f_n(x)\, t^n = \frac{1}{\sqrt{1-4t}}\left(\frac{2}{1+\sqrt{1-4t}}\right)^a F\left(\frac{-4xt}{[1+\sqrt{1-4t}]^2}\right),$$

where

$$F(u) = \sum_{k=0}^{\infty} \gamma_k u^k,$$

then

$$\sum_{n=0}^{\infty} \frac{a}{a+2n} f_n(x)\, t^n = \left(\frac{2}{1+\sqrt{1-4t}}\right)^a \psi\left(\frac{-4xt}{[1+\sqrt{1-4t}]^2}\right)$$

where

$$\psi(u) = \sum_{k=0}^{\infty} \frac{a}{a+2k} \gamma_k u^k.$$

Proof of Theorem 2: Let $G(x,t)$ denote the generating function of the hypothesis:

$$G(x,t)=\frac{1}{\sqrt{1-4t}}\left(\frac{2}{1+\sqrt{1-4t}}\right)^a F\left(\frac{-4xt}{[1+\sqrt{1-4t}]^2}\right)$$
$$=\sum_{k=0}^{\infty}\frac{1}{\sqrt{1-4t}}\left(\frac{2}{1+\sqrt{1-4t}}\right)^{a+2k}\gamma_k(-xt)^k.$$

By Rainville [1; p. 70, with $2\gamma-1=a+2k$], we write the identity

$$\frac{1}{\sqrt{1-4t}}\left(\frac{2}{1+\sqrt{1-4t}}\right)^{a+2k}={}_2F_1\left[\begin{matrix}\frac{a+2k+1}{2}, & \frac{a+2k+2}{2}; \\ & a+2k+1;\end{matrix}\quad 4t\right].$$

Then by substitution, we get

$$G(x,t)=\sum_{k=0}^{\infty}{}_2F_1\left[\begin{matrix}\frac{a+2k+1}{2}, & \frac{a+2k+2}{2}; \\ & a+2k+1;\end{matrix}\quad 4t\right]\gamma_k(-xt)^k.$$

After simplification we find that

$$G(x,t)=\sum_{n=0}^{\infty}\sum_{k=0}^{n}\frac{(a+1)_{2n}\gamma_k(-1)^k x^k}{(a+1)_{n+k}(n-k)!}t^n.$$

Therefore, from the generating relation of the hypothesis,

$$f_n(x)=\sum_{k=0}^{n}\frac{(a+1)_{2n}\gamma_k(-1)^k x^k}{(a+1)_{n+k}(n-k)!}.$$

By using this expansion of $f_n(x)$ in powers of x, we get

$$\sum_{n=0}^{\infty}\frac{a}{a+2n}f_n(x)\,t^n=\sum_{k=0}^{\infty}\gamma_k(-xt)^k\sum_{n=0}^{\infty}\frac{(a)_{2n+2k}\,t^n}{(a+1)_{n+2k}\,n!}.$$

We derive the following identity:

$$\frac{(a)_{2n+2k}}{(a+1)_{n+2k}}=\frac{(a)_{2k}(a+2k)_{2n}}{(a+1)_{2k}(a+1+2k)_n}$$
$$=\frac{a}{a+2k}\,\frac{2^{2n}\left(\frac{a+2k}{2}\right)_n\left(\frac{a+2k+1}{2}\right)_n}{(a+1+2k)_n}.$$

By substitution we get

$$\sum_{n=0}^{\infty} \frac{a}{a+2n} f_n(x)\, t^n = \sum_{k=0}^{\infty} \gamma_k (-x\,t)^k \frac{a}{a+2k} \sum_{n=0}^{\infty} \frac{\left(\frac{a+2k}{2}\right)_n \left(\frac{a+2k+1}{2}\right)_n (4t)^n}{(a+1+2k)_n\, n!}$$

$$= \sum_{k=0}^{\infty} \gamma_k (-x\,t)^k \frac{a}{a+2k}\, {}_2F_1\left[\begin{matrix} \frac{a+2k}{2}, \frac{a+2k+1}{2}; \\ a+2k+1; \end{matrix}\; 4t\right].$$

From Rainville [1; p. 70] we also have the identity

$${}_2F_1\left[\begin{matrix} \gamma, \gamma-\frac{1}{2}; \\ 2\gamma; \end{matrix}\; z\right] = \left(\frac{2}{1+\sqrt{1-z}}\right)^{2\gamma-1}.$$

Then with $2\gamma = a+2k+1$ and $z=4t$, we have

$${}_2F_1\left[\begin{matrix} \frac{a+2k+1}{2}, \frac{a+2k}{2}; \\ a+2k+1; \end{matrix}\; 4t\right] = \left(\frac{2}{1+\sqrt{1-4t}}\right)^{a+2k}.$$

Therefore, we may write

$$\sum_{n=0}^{\infty} \frac{a}{a+2n} f_n(x)\, t^n = \sum_{k=0}^{\infty} \gamma_k (-x\,t)^k \frac{a}{a+2k} \left(\frac{2}{1+\sqrt{1-4t}}\right)^{a+2k}$$

$$= \left(\frac{2}{1+\sqrt{1-4t}}\right)^{a} \sum_{k=0}^{\infty} \frac{a}{a+2k} \gamma_k \left(\frac{-4x\,t}{[1+\sqrt{1-4t}]^2}\right)^k,$$

or equivalently,

$$\sum_{n=0}^{\infty} \frac{a}{a+2n} f_n(x)\, t^n = \left(\frac{2}{1+\sqrt{1-4t}}\right)^{a} \psi\left(\frac{-4x\,t}{[1+\sqrt{1-4t}]^2}\right),$$

where

$$\psi(u) = \sum_{k=0}^{\infty} \frac{a}{a+2k} \gamma_k\, u^k. \quad \square$$

If in the hypothesis of the theorem

$$\gamma_k = \frac{(\alpha_1)_k (\alpha_2)_k \dots (\alpha_p)_k}{(\beta_1)_k (\beta_2)_k \dots (\beta_q)_k} \frac{1}{k!},$$

it follows that

$$F(u) = {}_pF_q\left[\begin{matrix} \alpha_1, \alpha_2, \dots, \alpha_p; \\ \beta_1, \beta_2, \dots, \beta_q; \end{matrix}\; u\right].$$

Then since

$$\frac{a}{a+2k}=\frac{\left(\frac{a}{2}\right)_k}{\left(\frac{a}{2}+1\right)_k},$$

we have

$$\psi(u)={}_{p+1}F_{q+1}\left[\begin{matrix}\frac{a}{2}, & \alpha_1,\alpha_2,\ldots,\alpha_p; \\ \frac{a}{2}+1, & \beta_1,\beta_2,\ldots,\beta_q;\end{matrix}\; u\right].$$

In particular for the set $\{L_n^{a+n}(x)\}$ we have, with $\gamma_k=\frac{1}{k!}$,

$$\sum_{n=0}^{\infty} L_n^{a+n}(x)\,t^n=\frac{1}{\sqrt{1-4t}}\left(\frac{2}{1+\sqrt{1-4t}}\right)^a {}_0F_0\left[\begin{matrix}-; \\ -;\end{matrix}\; \frac{-4xt}{(1+\sqrt{1-4t})^2}\right] \tag{11}$$

and

$$\sum_{n=0} \frac{a}{a+2n} L_n^{a+n}(x)\,t^n=\left(\frac{2}{1+\sqrt{1-4t}}\right)^a {}_1F_1\left[\begin{matrix}\frac{a}{2}; \\ \frac{a}{2}+1;\end{matrix}\; \frac{-4xt}{(1+\sqrt{1-4t})^2}\right]. \tag{12}$$

Since either differentiation or integration actually alters the generating function, the various forms of a generating relation provide us with a choice for a particular use. We conclude this section by giving an example of the usefulness of an integrated form. By using (12) above, Brown [1; pp. 69–72] was able to prove that

$$\frac{x^n}{n!}=\frac{a+2n}{a}\sum_{k=0}^{n}\frac{(a)_{n+k}(-1)^k}{(a+1)_{2k}(n-k)!}L_k^{a+k}(x).$$

5. Generating functions related by the Laplace transform. From a known generating relation it is sometimes possible to determine a second generating relation by means of the Laplace transform or its inverse. Suppose the set of functions $\{f_n(x)\}$ is generated as follows:

$$F(x,t)=\sum_{n=0}^{\infty} f_n(x)\,t^n.$$

If in this given generating relation we replace t by $1/s$ and also multiply both members by $1/s^a$, we get

$$\frac{1}{s^a}F\left(x,\frac{1}{s}\right)=\sum_{n=0}^{\infty} f_n(x)\frac{1}{s^{n+a}}, \qquad \text{where } a\geqq 0.$$

If, for some choice of a, the inverse Laplace transform of the left member exists, we then have

$$L^{-1}\left\{\frac{1}{s^a} F\left(x, \frac{1}{s}\right)\right\} = \sum_{n=0}^{\infty} f_n(x) \frac{t^{n+a-1}}{\Gamma(n+a)}.$$

Example 1. $a=1$.

The simple Laguerre polynomials $\{L_n(x)\}$ satisfy the following generating relation:

$$(1-t)^{-1} \exp\left\{\frac{-xt}{1-t}\right\} = \sum_{n=0}^{\infty} L_n(x)\, t^n. \tag{1}$$

If in (1) we replace t by $1/s$ and multiply both members by $1/s$, we get

$$\frac{1}{s}\left(1-\frac{1}{s}\right)^{-1} \exp\left\{\frac{-x/s}{1-1/s}\right\} = \sum_{n=0}^{\infty} L_n(x) \frac{1}{s^{n+1}}, \tag{2}$$

which implies

$$\frac{1}{s-1} \exp\left\{\frac{-x}{s-1}\right\} = \sum_{n=0}^{\infty} L_n(x) \frac{1}{s^{n+1}}. \tag{3}$$

Therefore, by finding the inverse transform of both members of (3), we find the generating relation

$$e^t\, {}_0F_1(-;\, 1;\, -xt) = \sum_{n=0}^{\infty} L_n(x) \frac{t^n}{n!}.$$

See Rainville [1; p. 213, (3)].

Example 2. $a=2\lambda$.

From the generating relation

$$\sum_{n=0}^{\infty} C_n^{\lambda}(x)\, z^n = (1-2xz+z^2)^{-\lambda}, \tag{1}$$

with x replaced by $\cos\theta$, we can derive the following generating relation by using the inverse Laplace transform:

$$\sum_{n=0}^{\infty} C_n^{\lambda}(\cos\theta) \frac{z^n}{(2\lambda)_n} = \Gamma(\lambda+\tfrac{1}{2})\, e^{z\cos\theta} (\tfrac{1}{2} z \sin\theta)^{\frac{1}{2}-\lambda} J_{\lambda-\frac{1}{2}}(z\sin\theta). \tag{2}$$

We note first that

$$(1-2z\cos\theta+z^2)^{-\lambda} = (1-e^{i\theta} z)^{-\lambda} (1-e^{-i\theta} z)^{-\lambda}.$$

If in (1) we replace z by $1/s$, we get

$$\sum_{n=0}^{\infty} C_n^{\lambda}(\cos\theta) \frac{1}{s^n} = \left(1-\frac{e^{i\theta}}{s}\right)^{-\lambda} \left(1-\frac{e^{-i\theta}}{s}\right)^{-\lambda}.$$

If we multiply both members by $1/s^{2\lambda}$, we get

$$\sum_{n=0}^{\infty} C_n^{\lambda}(\cos\theta)\frac{1}{s^{n+2\lambda}}=(s-e^{i\theta})^{-\lambda}(s-e^{-i\theta})^{-\lambda}.$$

By finding the inverse transform of both members, we have

$$\sum_{n=0}^{\infty} C_n^{\lambda}(\cos\theta)\frac{z^{n+2\lambda-1}}{\Gamma(n+2\lambda)}$$
$$=\frac{\pi^{\frac{1}{2}}}{\Gamma(\lambda)}\left(\frac{z}{-e^{i\theta}+e^{-i\theta}}\right)^{\lambda-\frac{1}{2}}\exp\{-\tfrac{1}{2}z(-e^{i\theta}-e^{-i\theta})\}\,I_{\lambda-\frac{1}{2}}\left(\frac{-e^{i\theta}+e^{-i\theta}}{2}z\right),$$

where $I_v(z)$ is the modified Bessel function of the first kind. Then

$$\sum_{n=0}^{\infty} C_n^{\lambda}(\cos\theta)\frac{z^n}{(2\lambda)_n}$$
$$=\frac{z^{-2\lambda+1}\Gamma(2\lambda)\Gamma(\frac{1}{2})}{\Gamma(\lambda)}(2i)^{-\lambda+\frac{1}{2}}\left(\frac{-z}{\sin\theta}\right)^{\lambda-\frac{1}{2}}\exp\{z\cos\theta\}\,i^{-\lambda+\frac{1}{2}} \tag{3}$$
$$\times J_{\lambda-\frac{1}{2}}(z\sin\theta),$$

since $I_v(z)=i^{-v}J_v(iz)$. By simplifying (3) we obtain the desired form (2). (See Erdélyi [2; p. 177, (29) and (30)].)

In the examples above we have used the *inverse* Laplace transform. On the other hand, if we can find the Laplace transform of a given generating function, we can obtain a second generating function. Suppose $G(x,t)$ generates the set $\left\{\frac{\phi_n(x)}{n!}\right\}$, i.e.,

$$G(x,t)=\sum_{n=0}^{\infty}\phi_n(x)\frac{t^n}{n!}.$$

If we can find the Laplace transform, we get

$$L\{G(x,t)\}=\sum_{n=0}^{\infty}\phi_n(x)\frac{1}{s^{n+1}}.$$

By multiplying both members by s, we have

$$s\,L\{G(x,t)\}=\sum_{n=0}^{\infty}\phi_n(x)\frac{1}{s^n}.$$

Let $H(x,t)=[s\,L\{G(x,t)\}]_{s=1/t}$. Then

$$H(x,t)=\sum_{n=0}^{\infty}\phi_n(x)\,t^n.$$

We illustrate this procedure with the following example.

Example 3.

From Rainville [1; p. 303, ex. 10], we have the following generating function for the Gottlieb polynomials:

$$e^t {}_1F_1(1+x;\,1;\,-t(1-e^{-\lambda}))=\sum_{n=0}^{\infty}\frac{\phi_n(x;\lambda)\,t^n}{n!}.$$

Then

$$L\{e^t {}_1F_1(1+x;\,1;\,-t(1-e^{-\lambda}))\}=\sum_{n=0}^{\infty}\phi_n(x;\lambda)\frac{1}{s^{n+1}},$$

or equivalently,

$$sL\sum_{k=0}^{\infty}\frac{(1+x)_k(-1)^k(1-e^{-\lambda})^k}{(k!)^2}(e^t t^k)=\sum_{n=0}^{\infty}\phi_n(x;\lambda)\frac{1}{s^n}.$$

Therefore,

$$s\sum_{k=0}^{\infty}\frac{(1+x)_k(-1)^k(1-e^{-\lambda})^k}{(k!)^2}\,\frac{\Gamma(k+1)}{(s-1)^{k+1}}=\sum_{n=0}^{\infty}\phi_n(x;\lambda)\frac{1}{s^n}.$$

If we now substitute t for $1/s$, we get

$$\frac{1}{t}\sum_{k=0}^{\infty}\frac{(1+x)_k(-1)^k(1-e^{-\lambda})^k}{k!}\,\frac{1}{\left(\frac{1}{t}-1\right)^{k+1}}=\sum_{n=0}^{\infty}\phi_n(x;\lambda)\,t^n.$$

When the left member is simplified, we then have

$$(1-t)^x(1-t\,e^{-\lambda})^{-x-1}=\sum_{n=0}^{\infty}\phi_n(x;\lambda)\,t^n.$$

See Gottlieb [1; p. 455].

6. The contour integral method. The contour integral method used by Brafman to obtain generating functions (see Brafman [1] and [2]) is applicable to sets of functions satisfying a Rodrigues-type formula reducible to the form

$$f_n(x)=\frac{1}{n!}\,D_x^n[(a\,x+b)^n\,F(x)], \tag{1}$$

where a and b are constants, not both zero, and $F(x)$ is independent of n and differentiable an arbitrary number of times. For example, the Rodrigues-type formula for the Laguerre polynomials, usually written in the form

$$L_n^{(\alpha)}(x)=\frac{e^x\,x^{-\alpha}}{n!}\,D_x^n[e^{-x}\,x^{n+\alpha}],$$

may be written as follows:

$$f_n(x) \equiv e^{-x} x^\alpha L_n^{(\alpha)}(x) = \frac{1}{n!} D_x^n [x^n (e^{-x} x^\alpha)], \tag{2}$$

where, in accordance with the notation of (1), we have

$$a=1, \quad b=0, \quad F(x)=e^{-x} x^\alpha.$$

By means of this method a known generating relation, containing at least one arbitrary parameter, may be transformed into another generating relation. Brafman used the generating relation

$$(1-t)^{-a} {}_1F_1 \left[\begin{matrix} a; & \dfrac{-xt}{1-t} \\ 1+\alpha; & \end{matrix} \right] = \sum_{n=0}^{\infty} \frac{(a)_n L_n^{(\alpha)}(x) t^n}{(1+\alpha)_n} \tag{3}$$

to obtain the bilateral generating relation

$$\begin{aligned} &(1-t)^{-1-\alpha+a}(1-t+vt)^{-a} \exp\left\{\frac{-xt}{1-t}\right\} {}_1F_1 \left[\begin{matrix} a; & \dfrac{vtx}{(1-t)(1-t+vt)} \\ 1+\alpha; & \end{matrix} \right] \\ &\quad = \sum_{n=0}^{\infty} {}_2F_1 \left[\begin{matrix} n, \quad a; & v \\ 1+\alpha; & \end{matrix} \right] L_n^{(\alpha)}(x) t^n. \end{aligned} \tag{4}$$

See Brafman [1; p. 180, (5)]. For $v=0$, we see that (4) reduces to the familiar generating relation

$$(1-t)^{-1-\alpha} \exp\left\{\frac{-xt}{1-t}\right\} = \sum_{n=0}^{\infty} L_n^{(\alpha)}(x) t^n. \tag{5}$$

In order to obtain (4) from (3) Brafman used the following result due to Chaundy (see Chaundy [1] or Erdélyi [3; p. 267, (22)]):

$$\begin{aligned} &(1-t)^{-1} {}_{p+1}F_q \left[\begin{matrix} 1, \alpha_1, \ldots, \alpha_p; & \dfrac{-xt}{1-t} \\ \beta_1, \beta_2, \ldots, \beta_q; & \end{matrix} \right] \\ &\quad = \sum_{n=0}^{\infty} {}_{p+1}F_q \left[\begin{matrix} -n, \alpha_1, \ldots, \alpha_p; & x \\ \beta_1, \beta_2, \ldots, \beta_q; & \end{matrix} \right] t^n. \end{aligned} \tag{6}$$

Using

$$\begin{aligned} &(1-t)^{-1} {}_{p+2}F_q \left[\begin{matrix} \tfrac{1}{2}, 1, \alpha_1, \ldots, \alpha_p; & \dfrac{xt^2}{(1-t)^2} \\ \beta_1, \beta_2, \ldots, \beta_q; & \end{matrix} \right] \\ &\quad \cong \sum_{n=0}^{\infty} {}_{p+2}F_q \left[\begin{matrix} -\tfrac{1}{2}n, -\tfrac{1}{2}n+\tfrac{1}{2}, \alpha_1, \ldots, \alpha_p; & x \\ \beta_1, \beta_2, \ldots, \beta_q; & \end{matrix} \right] t^n, \end{aligned} \tag{7}$$

instead of Chaundy's formula given in (6), Brafman obtained by the contour integral transformation of (3) another generating function for $\{L_n^{(\alpha)}(x)\}$:

$$\tfrac{1}{2}\exp\left\{\frac{-xt}{1-t}\right\}(1-t)^{a-\alpha-1}(1-t-vt)^{-a}{}_1F_1\left[\begin{matrix}a; \\ 1+\alpha;\end{matrix}\ \frac{-tvx}{(1-t)(1-t-vt)}\right]$$
$$+(1-t+vt)^{-a}{}_1F_1\left[\begin{matrix}a; \\ 1+\alpha;\end{matrix}\ \frac{tvx}{(1-t)(1-t+vt)}\right] \tag{8}$$
$$=\sum_{n=0}^{\infty}{}_4F_3\left[\begin{matrix}-\frac{1}{2}n,\ -\frac{1}{2}n+\frac{1}{2},\ \frac{1}{2}a,\ \frac{1}{2}a+\frac{1}{2}; \\ \frac{1}{2},\ \frac{1}{2}+\frac{1}{2}\alpha,\ 1+\frac{1}{2}\alpha;\end{matrix}\ v^2\right]L_n^{(\alpha)}(x)\,t^n.$$

See Brafman [1; p. 184, (28) and p. 180, (7)].

We will give the method in outline before applying it to a specific set of functions. The Rodrigues representation (1) is converted by means of the Cauchy integral theorem into the following contour integral form:

$$f_n(x)=\frac{1}{2\pi i}\int_C \frac{(az+b)^n F(z)}{(z-x)^{n+1}}\,dz, \tag{9}$$

where x is an interior point of the region bounded by the closed contour C. Furthermore, we require that $(az+b)^n F(z)$ be analytic within and on the closed contour C. (See Churchill [1; p. 156].)

If we want to use Chaundy's formula given in (6), we multiply both members of (9) by

$${}_{p+1}F_q\left[\begin{matrix}-n, & \alpha_1,\dots,\alpha_p; \\ & \beta_1,\dots,\beta_q;\end{matrix}\ v\right]t^n$$

and sum over n. If we interchange the operations of summation and integration, apply Chaundy's formula, and again interchange operations, we obtain a formal (not necessarily convergent) relation.

By using Chaundy's formula (6), we will obtain a bilateral generating relation for the set $\{f_n^a(x)\}$, where $f_n^a(x)=(-1)^n L_n^{-a-n}(x)$. We shall assume as given the following two relations

$$f_n^a(x)=\frac{(-1)^n}{n!}\,x^{a+n}e^x D^n[e^{-x}x^{-a}], \tag{10}$$

where $D=d/dx$, and

$$(1-xt)^{-c}{}_2F_0\left[\begin{matrix}c, & a; \\ & -;\end{matrix}\ \frac{t}{1-xt}\right]\cong\sum_{n=0}^{\infty}(c)_n f_n^a(x)\,t^n, \tag{11}$$

where c is a nonnegative real number. By means of Cauchy's integral formula we express (10) in the following form:

$$f_n^a(x)=\frac{(-1)^n}{n!}\,x^{a+n}e^x\frac{n!}{2\pi i}\int_C\frac{e^{-z}z^{-a}}{(z-x)^{n+1}}\,dz, \tag{12}$$

where C is any closed contour for which $z=0$ is an exterior point. If we multiply both members of (12) by

$$t^n \;{}_{p+1}F_q\left[\begin{matrix}-n, & \alpha_1,\dots,\alpha_p; \\ & \beta_1,\dots,\beta_q;\end{matrix}\; v\right]$$

and sum over n, we get

$$\begin{aligned}\sum_{n=0}^{\infty} {}_{p+1}F_q&\left[\begin{matrix}-n, & \alpha_1,\dots,\alpha_p; \\ & \beta_1,\dots,\beta_q;\end{matrix}\; v\right] f_n^a(x)\, t^n \\ &= \sum_{n=0}^{\infty} \frac{t^n(-1)^n x^{a+n} e^x}{2\pi i}\;{}_{p+1}F_q\left[\begin{matrix}-n, & \alpha_1,\dots,\alpha_p; \\ & \beta_1,\dots,\beta_q;\end{matrix}\; v\right] \int_C \frac{e^{-z} z^{-a}\, dz}{(z-x)^{n+1}}.\end{aligned} \tag{13}$$

Let $L(v,x,t)$ represent the left member of (13) and $R(v,x,t)$ represent the right member of (13). We propose to simplify $R(v,x,t)$ by formally interchanging the order of the operations of summation and integration and then applying Chaundy's formula:

$$\begin{aligned}R(v,x,t) \cong \frac{x^a e^x}{2\pi i} &\int_C \frac{e^{-z} z^{-a}}{z-x} \\ &\cdot \sum_{n=0}^{\infty} {}_{p+1}F_q\left[\begin{matrix}-n, & \alpha_1,\dots,\alpha_p; \\ & \beta_1,\dots,\beta_q;\end{matrix}\; v\right] \left(\frac{-x t}{z-x}\right)^n dz.\end{aligned} \tag{14}$$

In Chaundy's formula (6) we replace x by v and t by $(-xt)/(z-x)$ to get the following relation:

$$\begin{aligned}\sum_{n=0}^{\infty} {}_{p+1}F_q&\left[\begin{matrix}-n, & \alpha_1,\dots,\alpha_p; \\ & \beta_1,\dots,\beta_q;\end{matrix}\; v\right] \left(\frac{-x t}{z-x}\right)^n \\ &= \left(\frac{z-x-xt}{z-x}\right)^{-1} {}_{p+1}F_q\left[\begin{matrix}1, & \alpha_1,\dots,\alpha_p; \\ & \beta_1,\dots,\beta_q;\end{matrix}\; \frac{v x t}{z-x+xt}\right].\end{aligned} \tag{15}$$

If we substitute (15) in (14), we get

$$\begin{aligned}R(v,x,t) &\cong \frac{x^a e^x}{2\pi i} \int_C \frac{e^{-z} z^{-a}}{z-x} \left(\frac{z-x-xt}{z-x}\right)^{-1} \\ &\qquad \cdot {}_{p+1}F_q\left[\begin{matrix}1, & \alpha_1,\dots,\alpha_p; \\ & \beta_1,\dots,\beta_q;\end{matrix}\; \frac{v x t}{z-x-xt}\right] dz \\ &\cong \frac{x^a e^x}{2\pi i} \int_C \frac{e^{-z} z^{-a}}{z-x-xt} \sum_{n=0}^{\infty} \frac{(\alpha_1)_n \cdots (\alpha_p)_n}{(\beta_1)_n \cdots (\beta_q)_n} \left(\frac{v x t}{z-x-xt}\right)^n dz.\end{aligned}$$

We again interchange the order of operations and at the same time rearrange factors so that we may use (12), with x replaced by $x(1-t)$, to simplify the result:

$$R(v,x,t) \cong \sum_{n=0}^{\infty} \frac{(-1)^n [x(1-t)]^{a+n} e^{x(1-t)}}{2\pi i} \int_C \frac{e^{-z} z^{-a} dz}{[z-x(1-t)]^{n+1}} \times \frac{(-1)^n (1-t)^{-a-n} e^{xt} (\alpha_1)_n \cdots (\alpha_p)_n (vt)^n}{(\beta_1)_n \cdots (\beta_q)_n}.$$

Then by (12)

$$R(v,x,t) \cong \sum_{n=0}^{\infty} f_n^a(x(1-t)) \frac{e^{xt}}{(1-t)^a} \frac{(\alpha_1)_n \cdots (\alpha_p)_n}{(\beta_1)_n \cdots (\beta_q)_n} \left(\frac{-vt}{1-t}\right)^n. \tag{16}$$

Now from (13) and (16) we have

$$\sum_{n=0}^{\infty} {}_{p+1}F_q \left[\begin{matrix} -n, & \alpha_1, \dots, \alpha_p; \\ & \beta_1, \dots, \beta_q; \end{matrix} \; v \right] f_n^a(x) t^n \cong \frac{e^{xt}}{(1-t)^a} \sum_{n=0}^{\infty} \frac{(\alpha_1)_n \cdots (\alpha_p)_n}{(\beta_1)_n \cdots (\beta_p)_n} f_n^a(x(1-t)) \left(\frac{-vt}{1-t}\right)^n.$$

In order to convert the right member of this general relation into a specific generating function, we must have available a function which generates $\{f_n^a(x)\}$ and which contains at least one arbitrary constant. The generating function of (11), with arbitrary constant c, satisfies these requirements.

In order to use (11) we choose $p=1$, $\alpha_1=c$, $q=0$. For this choice

$$L(v,x,t) \cong \sum_{n=0}^{\infty} {}_2F_0 \left[\begin{matrix} -n, & c; \\ & -; \end{matrix} \; v \right] f_n^a(x) t^n, \tag{17}$$

and from (16)

$$R(v,x,t) \cong \frac{e^{xt}}{(1-t)^a} \sum_{n=0}^{\infty} (c)_n f_n^a(x(1-t)) \left(\frac{-vt}{1-t}\right)^n. \tag{18}$$

If in (11) we replace x by $x(1-t)$ and t by $(-vt)/(1-t)$, we get

$$(1+vxt)^{-c} {}_2F_0 \left[\begin{matrix} c, & a; \\ & -; \end{matrix} \; \frac{-vt}{(1-t)(1+vxt)} \right] \cong \sum_{n=0}^{\infty} (c)_n f_n^a(x(1-t)) \left(\frac{-vt}{1-t}\right)^n.$$

Then from (18)

$$R(v,x,t) \cong \frac{e^{xt}}{(1-t)^a} (1+vxt)^{-c} {}_2F_0 \left[\begin{matrix} c, & a; \\ & -; \end{matrix} \; \frac{-vt}{(1-t)(1+vxt)} \right]. \tag{19}$$

Finally from (17) and (19) we obtain the (divergent) generating relation

$$\sum_{n=0}^{\infty} {}_2F_0\left[\begin{matrix}-n, & c; \\ & -;\end{matrix}\; v\right] f_n^a(x)\, t^n$$
$$\cong e^{xt}(1-t)^{-a}(1+v\,x\,t)^{-c}\, {}_2F_0\left[\begin{matrix}c, & a; \\ & -;\end{matrix}\; \frac{-v\,t}{(1-t)(1+v\,x\,t)}\right]. \tag{20}$$

The bilateral generating function of (20) was obtained in Chapter I, Section 6, by series manipulation and in Chapter III, Section 2, by Weisner's method.

7. Recent developments. During the five-year period from 1965 through 1969, mathematicians throughout the world have continued to be interested in generating functions. This interest is evidenced by publications coming from India, Japan, Canada, the United States, England, Russia, Germany, and other European countries.

In 1968 alone three outstanding books on special functions were published. Each of these considered the generating function concept from a group-theoretic viewpoint. They were written by Willard Miller, Jr., [1] of the United States, by James D. Talman [1] of Canada, and by N. J. Vilenkin [1] of Russia. (Vilenkin's book was translated into English by V. N. Singh.)

Among the many interesting articles written on generating functions and published recently are articles by N. A. Al-Salam and W. A. Al-Salam [1], by J. W. Brown [2], and by L. Carlitz [1]. The reader will find numerous references to articles on generating functions in the Special Functions Section of the *Mathematical Reviews*.

Bibliography

Al-Salam, N.A., Al-Salam, W.A.: [1] Some characterizations of the ultraspherical polynomials. Canad. Math. Bull. **11**, 457–464 (1968).

Bedient, Phillip E.: [1] Polynomials related to Appell functions of two variables. Michigan thesis, 1958.

Bell, Eric T.: [1] Euler algebra. Trans. Amer. Math. Soc. **25**, 135–154 (1923).

— [2] Algebraic arithmetic, vol. 7. Providence: Amer. Math. Soc. Colloquium Publications 1927.

— [3] Postulational bases for the umbral calculus. Amer. J. Math. **62**, 717–724 (1940).

Bessel, Friedrich Wilhelm: [1] Untersuchung des Theils der planetarischen Störungen, welcher aus der Bewegung der Sonne entsteht. Abh. Akad. Wiss. Berlin, Kl. Math. (1824) or Abhandlungen von Friedrich Wilhelm Bessel, I, 16, S. 84–109. Leipzig: Wilhelm Engelmann 1875.

Boas, Ralph P., Jr., Buck, R. Creighton: [1] Polynomial expansions of analytic functions. Berlin-Göttingen-Heidelberg: Springer 1958.

— [2] Polynomials defined by generating relations. Amer. Math. Monthly **63**, 626–632 (1956).

Brafman, Fred: [1] Some generating functions for Laguerre and Hermite polynomials. Canad. J. Math. **9**, 180–187 (1957).

— [2] Generating functions and Associated Legendre polynomials. Quart. J. Math. Oxford Ser. **10**, 156–160 (1959).

— [3] Generating functions of Jacobi and related polynomials. Proc. Amer. Math. Soc. **2**, 942–949 (1951).

Brown, James Ward: [1] On certain modifications of Gegenbauer and Laguerre polynomials. Michigan thesis, 1964.

— [2] On zero type sets of Laguerre polynomials. Duke Math. J. **35**, 821–823 (1968).

Burchnall, J.L.: [1] The Bessel polynomials. Canad. J. Math. **3**, 62–68 (1951).

Carlitz, Leonard: [1] Some generating functions for Laguerre polynomials. Duke Math. J. **35**, 825–827 (1968).

Chaundy, T.W.: [1] An extension of the hypergeometric function. Quart. J. Math. Oxford Ser. **14**, 55–78 (1943).

Churchill, Ruel V.: [1] Operational mathematics, 2nd, ed. New York: McGraw-Hill 1958.

Cohn, P.M.: [1] Lie groups. Cambridge: University Press 1961.

Erdélyi, Arthur, with Magnus, W., Oberhettinger, F., Tricomi, F.G., et al.: [1] Higher transcendental functions, vol. 1. New York: McGraw-Hill 1953.

— [2] Higher transcendental function, vol. 2. New York: McGraw-Hill 1953.

— [3] Higher transcendental function, vol. 3. New York: McGraw-Hill 1955.

Gegenbauer, Leopold: [1] Über die Bessel'schen Funktionen. Akad. Wiss. Wien, Abteilung II a, **70**, 6–16 (1874).

Gottlieb, Morris J.: [1] Concerning some polynomials orthogonal on a finite or enumerable set of points. Amer. J. Math. **60**, 453–458 (1938).

Hermite, Charles: [1] Sur un nouveau développement en série de fonctions. C. r. Acad. Sci., Paris **58**, 93–100, 266–273 (1864) or Oeuvres, vol. 2, p. 293–312. Paris: Gauthier-Villars 1908.

Hochstadt, Harry: [1] Special functions of mathematical physics. New York: Holt, Rinehart and Winston 1961.

Jordan, Charles: [1] Calculus of finite differences. New York: Chelsea 1947.

Krall, H.L., Frink, Orrin: [1] A new class of orthogonal polynomials: the Bessel polynomials. Trans. Amer. Math. Soc. **65**, 100-115 (1949).

Laguerre, Edmond Nicolas: [1] Sur l'integrale $\int_x^\infty \frac{e^{-x}dx}{x}$. Bull. Soc. Math. France **7**, 72-81 (1879) or Oeuvres, vol. 1, p. 428-437. Paris: Gauthier-Villars 1898.

Legendre, Adrian Marie: [1] Recherches sur l'attraction des sphéroïdes homogènes. Mém. math. phys. prés a l'Acad. Sc. **10**, 411-434 (1785).

Lie, Sophus: [1] Gesammelte Abhandlungen, vol. 3. Leipzig: B. G. Teubner 1922.

Magnus, Wilhelm, Oberhettinger, Fritz: [1] Formulas and theorems for the special functions of mathematical physics. New York: Chelsea 1949.

Meixner, Josef: [1] Erzeugende Funktionen der Charlierschen Polynome. Math. Z. **44**, 531-535 (1939).

Miller, Frederic H.: [1] Partial differential equations. New York: John Wiley & Sons 1941.

Miller, Willard, Jr.: [1] Lie theory and special functions. New York: Academic Press 1968.

Milne-Thomson, L.M.: [1] The calculus of finite differences. London: Macmillan 1933.

Morse, Philip M., Feshbach, Herman: [1] Methods of theoretical physics, vol. 1. New York: McGraw-Hill 1953.

— [2] Methods of theoretical physics, vol. 2. New York: McGraw-Hill 1953.

Obrechkoff, Nikola: [1] Sur la summation des séries trigonométriques de Fourier par les moyennes arithmetiques. Bull. Soc. Math. France **62**, 84-109 (1934).

Rainville, Earl D.: [1] Special functions. New York: Macmillan 1960.

Sheffer, I.M.: [1] Some properties of polynomial sets of type zero. Duke Math. J. **5**, 590-622 (1939).

Sneddon, Ian N.: [1] Special functions of mathematical physics and chemistry. New York: Interscience 1961.

Talman, James D.: [1] Special functions. New York: W. A. Benjamin 1968.

Truesdell, Clifford A.: [1] A unified theory of special functions. Princetown: University Press 1948.

— [2] On the addition and multiplication theorems for special functions. Proc. Nat. Acad. Sci. U.S.A. **36**, 752-755 (1950).

— [3] A note on the Poisson-Charlier functions. Ann. Math. Statist. **18**, 450-454 (1947).

Vilenkin, N.Ja.: [1] Special functions and the theory of group representations. Providence: Amer. Math. Soc. Transl. 1968.

Viswanathan, Bhaskaran: [1] Generating functions for ultraspherical functions. Canad. J. Math. **20**, 120-134 (1968).

Watson, George Neville: [1] A treatise on the theory of Bessel functions, 2nd ed. Cambridge: University Press 1944.

Weisner, Louis: [1] Group-theoretic origins of certain generating functions. Pacific J. Math. **5**, 1033-1039 (1955).

— [2] Generating functions for Hermite functions. Canad. J. Math. **11**, 141-147 (1959).

— [3] Generating functions for Bessel functions. Canad. J. Math. **11**, 148-155 (1959).

Whittaker, E.T., Watson, G.N.: [1] Modern analysis, 4th ed. Cambridge: University Press 1927.

Index

Springer Tracts in Natural Philosophy

Vol. 1 **Gundersen: Linearized Analysis of One-Dimensional Magnetohydrodynamic Flows**
With 10 figures. X, 119 pages. 1964. Cloth DM 22,—; US $ 6.10

Vol. 2 **Walter: Differential- und Integral-Ungleichungen** und ihre Anwendung bei Abschätzungs- und Eindeutigkeitsproblemen
Mit 18 Abbildungen. XIV, 269 Seiten. 1964. Gebunden DM 59,—; US $ 16.30

Vol. 3 **Gaier: Konstruktive Methoden der konformen Abbildung**
Mit 20 Abbildungen und 28 Tabellen. XIV, 294 Seiten. 1964.
Gebunden DM 68,—; US $ 18.70

Vol. 4 **Meinardus: Approximation von Funktionen und ihre numerische Behandlung**
Mit 21 Abbildungen. VIII, 180 Seiten. 1964. Gebunden DM 49,—; US $ 13.50

Vol. 5 **Coleman, Markovitz, Noll: Viscometric Flows of Non-Newtonian Fluids.**
Theory and Experiment
With 37 figures. XII, 130 pages. 1966. Cloth DM 22,—; US $ 6.10

Vol. 6 **Eckhaus: Studies in Non-Linear Stability Theory**
With 12 figures. VIII, 117 pages. 1965. Cloth DM 22,—; US $ 6.10

Vol. 7 **Leimanis: The General Problem of the Motion of Coupled Rigid Bodies About a Fixed Point**
With 66 figures. XVI, 337 pages. 1965. Cloth DM 48,—; US $ 13.20

Vol. 8 **Roseau: Vibrations non linéaires et théorie de la stabilité**
Avec 7 figures. XII, 254 pages. 1966. Relié DM 39,—; US $ 10.80

Vol. 9 **Brown: Magnetoelastic Interactions**
With 13 figures. VIII, 155 pages. 1966. Cloth DM 38,—; US $ 10.00

Vol. 10 **Bunge: Foundations of Physics**
With 5 figures. XII, 311 pages. 1967. Cloth DM 56,—; US $ 14.80

Vol. 11 **Lavrentiev: Some Improperly Posed Problems of Mathematical Physics**
With 1 figure. VIII, 72 pages. 1967. Cloth DM 20,—; US $ 5.50

Vol. 12 **Kronmüller: Nachwirkung in Ferromagnetika**
Mit 92 Abbildungen. XIV, 329 Seiten. 1968. Gebunden DM 62,—; US $ 17.10

Vol. 13 **Meinardus: Approximation of Functions: Theory and Numerical Methods**
With 21 figures. VIII, 198 pages. 1967. Cloth DM 54,—; US $ 14.80

Vol. 14 **Bell: The Physics of Large Deformation of Crystalline Solids**
With 166 figures. X, 253 pages. 1968. Cloth DM 48,—; US $ 12.00

Vol. 15 **Buchholz: The Confluent Hypergeometric Function** with Special Emphasis on its Applications
XVIII, 238 pages. 1969. Cloth DM 64,—; US $ 16.00

Vol. 16 **Slepian: Mathematical Foundations of Network Analysis**
XI, 195 pages. 1968. Cloth DM 44,—; US $ 12.10

Vol. 17 **Gavalas: Nonlinear Differential Equations of Chemically Reacting Systems**
With 10 figures. IX, 107 pages. 1968. Cloth DM 34,—; US $ 9.40

Vol. 18 **Marti: Introduction to the Theory of Bases**
XII, 150 pages. 1969. Cloth DM 32,—; US $ 8.80

Vol. 19 **Knops, Payne: Uniqueness Theorems in Linear Elasticity**
Approx. 145 pages. Due May 1971. Cloth DM 36,—; US $ 9.90

Vol. 20 **Edelen, Wilson: Relativity and the Question of Discretization in Astronomy**
With 34 figures. XII, 186 pages. 1970. Cloth DM 38,—; US $ 10.50

Vol. 21 **McBride: Obtaining Generating Functions**
VIII, 100 pages. 1971. Cloth DM 44,—; US $ 12.10